NOMENCLATURE OF INORGANIC CHEMISTRY

IUPAC PERIODIC TABLE OF THE ELEMENTS

1	2	3	4	5	6	7	8	9	10	11	12	13	14	15	16	17	18	n
1 H																	2 He	1
3 Li	4 Be											5 B	6 C	7 N	8 O	9 F	10 Ne	2
11 Na	12 Mg											13 Al	14 Si	15 P	16 S	17 Cl	18 Ar	3
19 K	20 Ca	21 Sc	22 Ti	23 V	24 Cr	25 Mn	26 Fe	27 Co	28 Ni	29 Cu	30 Zn	31 Ga	32 Ge	33 As	34 Se	35 Br	36 Kr	4
37 Rb	38 Sr	39 Y	40 Zr	41 Nb	42 Mo	43 Tc	44 Ru	45 Rh	46 Pd	47 Ag	48 Cd	49 In	50 Sn	51 Sb	52 Te	53 I	54 Xe	5
55 Cs	56 Ba	57–71 La–Lu	72 Hf	73 Ta	74 W	75 Re	76 Os	77 Ir	78 Pt	79 Au	80 Hg	81 Tl	82 Pb	83 Bi	84 Po	85 At	86 Rn	6
87 Fr	88 Ra	89 103 Ac–Lr	104 Unq	105 Unp	106 Unh	107 Uns	108 Uno	109 Une	110 Uun	111 Uuu	112 Uub	113 Uut	114 Uuq	115 Uup	116 Uuh	117 Uus	118 Uuo	7

															n
57 La	58 Ce	59 Pr	60 Nd	61 Pm	62 Sm	63 Eu	64 Gd	65 Tb	66 Dy	67 Ho	68 Er	69 Tm	70 Yb	71 Lu	6
89 Ac	90 Th	91 Pa	92 U	93 Np	94 Pu	95 Am	96 Cm	97 Bk	98 Cf	99 Es	100 Fm	101 Md	102 No	103 Lr	7

International Union of Pure and Applied Chemistry

Nomenclature of Inorganic Chemistry
RECOMMENDATIONS 1990

Issued by the Commission on the
Nomenclature of Inorganic Chemistry
and edited by G. J. Leigh

OXFORD

BLACKWELL SCIENTIFIC PUBLICATIONS

LONDON EDINBURGH BOSTON

MELBOURNE PARIS BERLIN VIENNA

© 1990 International Union of
Pure and Applied Chemistry and
published for them by
Blackwell Scientific Publications
Editorial offices:
Osney Mead, Oxford OX2 0EL
25 John Street, London WC1N 2BL
23 Ainslie Place, Edinburgh EH3 6AJ
238 Main Street, Cambridge
    Massachusetts 02142, USA
54 University Street, Carlton
    Victoria 3053, Australia

Other Editorial Offices:
Librairie Arnette SA
2, rue Casimir-Delavigne
75006 Paris
France

Blackwell Wissenschafts-Verlag
Meinekestrasse 4
D-1000 Berlin 15
West Germany

Blackwell MZV
Feldgasse 13
A-1238 Wien
Austria

First published 1990
Reprinted (with corrections) 1991, 1992

Set by Macmillan India Limited
Printed and bound in Great Britain
by Wm Clowes Ltd. Beccles

DISTRIBUTORS

Marston Book Services Ltd
PO Box 87
Oxford OX2 0DT
(*Orders*: Tel: 0865 791155
        Fax: 0865 791927
        Telex: 837515)

USA
Blackwell Scientific Publications, Inc.
238 Main Street
Cambridge, MA 02142
(*Orders:* Tel: 800 759-6102
        617 876-7000)

Canada
Oxford University Press
70 Wynford Drive
Don Mills
Ontario M3C 1J9
(*Orders*: Tel: 416 441-2941)

Australia
Blackwell Scientific Publications
(Australia) Pty Ltd
54 University Street
Carlton, Victoria 3053
(*Orders*: Tel: 03 347-0300)

British Library
Cataloguing in Publication Data

Nomenclature of inorganic chemistry.
    Recommendations, 1990/issued by the
    Commission on the
    Nomenclature of Inorganic Chemistry,
    International Union of Pure and
    Applied Chemistry.
    edited by G. J. Leigh.
    Bibliography: p.
    Includes index.
    ISBN 0-632-02319-8
    ISBN 0-632-02494-1 (pbk)
    1. Chemistry, Inorganic—Nomenclature.
    I. Leigh, G. J. II. International Union of
    Pure and Applied Chemistry. Commission on
    the Nomenclature of Inorganic Chemistry.
    QD149.N66 1990
    546′.014—dc19

Contents

CONTENTS

CONTENTS

CONTENTS

IUPAC Commission on the Nomenclature of Inorganic Chemistry

The membership of the Commission during the period 1971–1987 in which the present edition was prepared was as follows:

Chairmen

W. C. Fernelius (U.S.A.)	1971–1975
J. Chatt (U.K.)	1975–1981
Y. Jeannin (France)	1981–1985
D. H. Busch (U.S.A.)	1985–

Vice-Chairmen

K. A. Jensen (Denmark)	1949–1973
Y. Jeannin (France)	1975–1981
D. H. Busch (U.S.A.)	1981–1985
E. Fluck (F.R.G.)	1985–1987

Secretaries

J. E. Prue (U.K.)	1963–1972
D. M. P. Mingos (U.K.)	1973–1978
T. D. Coyle (U.S.A.)	1979–1982
E. Fluck (F.R.G.)	1983–1985
E. Samuel (France)	1985–

Ordinary titular members

R. M. Adams (U.S.A.)	1967–1975
L. F. Bertello (Argentina)	1971–1979
D. H. Busch (U.S.A.)	1979–1981
C. K. Buschbeck (F.R.G.)	1971–1979
J. Chatt (U.K.)	1959–1973
E. Fluck (F.R.G.)	1979–1985
P. Fodor-Csányi (Hungary)	1979–1987
Y. Jeannin (France)	1971–1975
R. S. Laitinen (Finland)	1985–
G. J. Leigh (U.K.)	1973–1985
B. F. Myasoedov (U.S.S.R.)	1971–1979
J. F. Nixon (U.K.)	1985–
W. H. Powell (U.S.A.)	1975–1979
J. Reedijk (Netherlands)	1979–1987
E. Samuel (France)	1981–1985
T. Sloan (U.S.A.)	1985–

Associate members

R. M. Adams (U.S.A.)	1975–1981
G. B. Bokij (U.S.S.R.)	1979–1983
M. W. G. de Bolster (Netherlands)	1983–1987
D. H. Busch (U.S.A.)	1977–1979
C. K. Buschbeck (F.R.G.)	1979–1985
J. Chatt (U.K.)	1973–1975
J. A. Connor (U.K.)	1981–1985
T. D. Coyle (U.S.A.)	1977–1979
T. Erdey-Gruz (Hungary)	1969–1977
W. C. Fernelius (U.S.A.)	1975–1979
E. Fluck (F.R.G.)	1977–1979
E. W. Godly (U.K.)	1979–1987
A. K. Holliday (U.K.)	1971–1973
K. A. Jensen (Denmark)	1973–1981
J. Klikorka (Czechoslovakia)	1974–1981
R. Laitinen (Finland)	1981–1985
G. J. Leigh (U.K.)	1971–1973
R. C. Mehrotra (India)	1981–
R. Metselaar (Netherlands)	1985–
J. F. Nixon (U.K.)	1981–1985
R. Poilblanc (France)	1985–
W. H. Powell (U.S.A.)	1969–1975
J. Reedijk (Netherlands)	1977–1979
J. Riess (France)	1973–1977
A. Romao-Dias (Portugal)	1979–1983
E. Samuel (France)	1977–1981
K. Samuelsson (Sweden)	1981–
A. M. Sargeson (Australia)	1985–
C. Schäffer (Denmark)	1971–1981
T. Sloan (U.S.A.)	1979–1985
A. Vlcek (Czechoslovakia)	1969–1977
E. Weiss (F.R.G.)	1969–1973
K. Wieghardt (F.R.G.)	1985–
K. Yamasaki (Japan)	1969–1977

National Representatives

M. O. Albers (R.S.A.)	1984–1987
P. J. Aymonio (Argentina)	1985–1987
L. F. Bertello (Argentina)	1979–1985
T. D. Coyle (U.S.A.)	1974–1977
H. H. Emons (G.D.R.)	1978–1983
S. Fallab (Switzerland)	1984–1987
E. Fluck (F.R.G.)	1975–1977
P. Fodor-Csányi (Hungary)	1977–1979
E. W. Godly (U.K.)	1978–1979
L. Y. Goh (Malaysia)	1983–1987

COMMISSION ON INORGANIC CHEMISTRY NOMENCLATURE

Principal Authors of this Edition

D. H. BUSCH
University of Kansas, Lawrence, USA

J. CHATT
University of Sussex, Brighton, UK

T. D. COYLE
National Bureau of Standards, Washington, DC, USA

W. C. FERNELIUS
Kent State University, Kent, Ohio, USA

P. FODOR-CSÁNYI
Eötvös Loránd University, Budapest, Hungary

E. W. GODLY
Laboratory of the Government Chemist, London, UK

Y. JEANNIN
Université Pierre et Marie Curie, Paris, France

J. B. LEACH
Oxford Polytechnic, Oxford, UK

G. J. LEIGH
AFRC IPSR, Brighton, UK

R. METSELAAR
Technische Hogeschool, Eindhoven, Netherlands

J. REEDIJK
Leiden University, Leiden, Netherlands

E. SAMUEL
Ecole Nationale Supérieure de Chimie, Paris, France

T. SLOAN
Chemical Abstracts Service, Columbus, Ohio, USA

Preface to the First Edition

In addition to members of the Commission on Inorganic Chemical Nomenclature, the present revision is the evolved work of various individuals who have served as regular members of the Commission since the '1940 Rules' appeared. Their names are listed in the volumes of Comptes Rendus, IUPAC, which have appeared since 1940.

Acknowledgement is also made of the co-operation of delegate and advisory members of the Commission, of members of nomenclature committees in a number of nations; also of Dr E. J. Crane, Editor of *Chemical Abstracts*.

The final editing of the 1957 Report is the work of a sub-committee, Prof. K. A. Jensen, Chairman, Prof. J. Bénard, Prof. A. Ölander and Prof. H. Remy.

1 November 1958

ALEXANDER SILVERMAN
Chairman

Preface to the Second Edition

The IUPAC Commission on the Nomenclature of Inorganic Chemistry, in its first meeting after the publication of the 1957 Rules (Munich 1959), scheduled further work for the Commission to deal with the nomenclature of boron hydrides and higher hydrides of the Group IV–VI elements, polyacids, and organometallic compounds. Part of this work, dealing with organometallic compounds, organoboron, organosilicon, and organophosphorus compounds, has been carried out in collaboration with the IUPAC Commission on the Nomenclature of Organic Chemistry. It has now been completed and Tentative Rules for this field will be published.

In the meantime work on a revision of the 1957 Rules has been going on continuously. Tentative proposals for changes or additions to this Report have been published in the Comptes Rendus of the Conferences in London (1963) and Paris (1965) and in the *IUPAC Information Bulletin*. However, as a result of comments to these tentative proposals some of them (such as the proposal to change chloro to chlorido) have not been retained in the final version. The section on coordination compounds has been much extended, reflecting the importance of this field in modern inorganic chemistry. A short section on boron hydrides and their derivatives has been included in the present edition, but a more extensive treatment has been published separately on a tentative basis in the *IUPAC Information Bulletin*: Appendices on Tentative Nomenclature, Symbols, Units and Standards, No. 8 (September 1970).

An editorial committee consisting of Prof. R. M. Adams, Prof. J. Chatt, Prof. W. C. Fernelius, Prof. F. Gallais, Dr. W. H. Powell, and Dr. J. E. Prue met in Columbus, Ohio, in the last week of January 1970, to finalize the manuscript. The Commission acknowledges the help of Dr. Kurt Loening, Chairman of the Commission on the Nomenclature of Macromolecular Chemistry, during this work. The work of the Commission was aided significantly by Grant No. 890-65 from the Air Force Office of Scientific Research administered by the U.S.A. National Academy of Sciences – National Research Council.

Copenhagen
9 September 1970

K. A. JENSEN
Chairman
Commission on the Nomenclature
of Inorganic Chemistry

Preface to the Current Edition

Chemical nomenclature is an ancient subject. The Greeks named ores or minerals and the products extracted from them and alchemists also gave names and symbols to their compounds. Later Lavoisier proposed the first systematic nomenclature for oxoacids. Chemistry develops and changes but inorganic chemical nomenclature is always a timely subject.

IUPAC has the responsibility to compile rules for nomenclature for the chemical community. The Rules for Inorganic Nomenclature were first formulated before the Second World War; they were reconsidered and discussed in the 1950s and published in 1959. A second revision took place in the 1960s and a second edition was published in 1971. Since then many new compounds have been prepared, some of which are difficult to name with the existing (1970) set of rules; this is because new types of bonding and new types of structure have appeared. Consequently, trivial and local systems of nomenclature have grown up, which emphasize the need for systematic, widely comprehensible nomenclature. This is particularly true in coordination chemistry and in boron chemistry. In 1978, the IUPAC Commission of Nomenclature on Inorganic Chemistry (CNIC) decided to replace the 1970 edition of *Nomenclature of Inorganic Chemistry* (the 'Red Book'). Because many of the new fields of chemistry are very highly specialized and need complex types of name, the new Red Book will appear in several parts. Part I is mainly concerned with what could be called the fundamental areas of inorganic chemistry. Subsequent parts will deal with specialized areas; typical examples are quasi-single strand inorganic polymers and polyoxoanions.

This present book is Part I. In it can be found rules to name compounds ranging from the simplest molecules to oxoacids and their derivatives, coordination compounds, and simple boron compounds.

It took ten years for CNIC to complete Part I. The individual Chapters were written by nominated individuals, assisted by working parties, and repeatedly reviewed and edited within CNIC. It was a considerable task to ensure consistency between the various Chapters dealing with so many different parts of chemistry. The aim of the Commission was to bring together established and traditional practices and systems of nomenclature so that they could be generally applicable. Where necessary, as in Chapters 6 and 11, the Commission leant heavily on outside expertise, and we acknowledge particularly the help of the Commission on High Temperature and Solid State Chemistry in the preparation of Chapter I-6. When a Chapter had been finished it was sent to external reviewers carefully chosen for their expertise in the field. Finally, the individual Chapters were submitted for discussion and criticism from the chemistry community by the usual IUPAC mechanism. These procedures should guarantee a good product which is generally acceptable.

We hope that this new edition will be a success and will be very widely used. The next edition is many years ahead and its production will be complicated because the subject will continue to evolve, and many of the new structures will not be amenable to treatment by current rules. It is our intention that the level of Part I should be such that its general principles will not be undermined, and that it should retain its currency for many years. Part II, more specialized, will appear in the near future, and, by its nature, will probably require more frequent revision.

The complete list of authors of all eleven Chapters is appended on p. xii. They are listed in alphabetical order without reference to their individual Chapters. The members of CNIC during the period 1971–1987 all contributed in various ways, even those who are not named specifically as authors. It is also in order to thank all the reviewers, too numerous to name because there were about fifteen for each Chapter. Finally, we wish to acknowledge the help of all those who submitted comments during the public review period. If we failed to thank any specific individuals formally, we hope that it will be satisfactory to acknowledge here all those who contributed to the genesis and production of this book.

23 February 1988

Y. Jeannin
Chairman
CNIC 1981–1985

Introduction to the First Edition (1957 Rules)

The Commission on the Nomenclature of Inorganic Chemistry of the International Union of Pure and Applied Chemistry (IUPAC) was formed in 1921, and many meetings took place which culminated in the drafting of a comprehensive set of Rules in 1938. On account of the war they were published in 1940 without outside discussion. At the meeting of the International Union of Chemistry in 1947 it was decided to undertake a thorough revision of what have come to be known as the '1940 Rules', and after much discussion they were completely rewritten and issued in English and French, the official languages of the Union, after the meeting in Stockholm in 1953 as the 'Tentative Rules for Inorganic Chemical Nomenclature'. These were studied by the various National Organizations and the comments and criticisms of many bodies and of private individuals were received and considered in Zürich, Switzerland, in 1955, in Reading, England, in 1956, and in Paris, France, in 1957.

The Rules set out here express the opinion of the Commission* as to the best general system of nomenclature, although certain names and rules which are given here as a basis for uniformity will probably prove unworkable or unacceptable in some languages. In these cases adaptation or even alteration will be necessary, but it is hoped that it will be possible to keep these changes small and to preserve the spirit of the IUPAC Rules. The English and French versions, which differ slightly, are to be regarded as international models from which translations will be made into other languages. The latter is likely to prove the better model for the Romance languages, and the former for Germanic languages, but it must be borne in mind that these languages are here used as the official languages of the Union and that several nations speak them with quite considerable variations of usage among themselves. There may therefore arise a similar need for adaptation or alteration even among English-speaking and French-speaking peoples, but we hope that in these cases, as in the others, careful consideration will be given to minimizing variation and to preserving the spirit of our international model.

The Commission's aim has been always to produce rules which lead to clear and acceptable names for as many inorganic compounds as possible. It soon became obvious, however, that different users may require the name of a compound to fulfil different objects, and it has been necessary to effect compromises in order to formulate rules of the most general utility. The principal function of a name is to provide the chemist with a word or set of words which is unique to the particular compound, and which conveys at least its empirical formula and also if possible its principal structural features. The name should be pronounceable and capable of being written or printed with an absolute minimum of additional symbols or modes of writing (*e.g.*, subscripts or differing type-faces).

---

* Chairman (1947–53) H. Bassett; (1953–57) Alex Silverman; Vice-Chairman, K. A. Jensen; Secretary, G. H. Cheesman; Members, J. Bénard, N. Bjerrum, E. H. Büchner, W. Feitknecht, L. Malatesta, A. Ölander, and H. Remy.

Many inorganic compounds exist only in the solid state, and are destroyed on fusion, solution or vaporization; some chemists have expressed strongly the view that names for such compounds should include information about the structure of the solid as well as its composition. Incorporating all this information tends to make the names extremely cumbersome, and since many of the structures remain uncertain or controversial, the names themselves are apt to be unstable. For general purposes, therefore, a considerable curtailment is essential and the Commission has endeavoured to produce a system based on the composition and most obvious properties of substances, avoiding as far as possible theoretical matters which are liable to change.

Introduction to the Second Edition
(1970 Rules)

A major revision and extension of Section 7 has been undertaken. The principle of an alphabetical order of citation of ligands in coordination entities has been adopted, and the rules now make detailed provision for the naming of complexes with unsaturated molecules or groups, the designation of ligand positions in the coordination sphere, the nomenclature of polynuclear compounds and those with metal–metal bonds, and the nomenclature of absolute configurations for six-coordinated complexes based on the octahedron. The former Section 4 which dealt with crystalline phases of variable composition has been similarly revised and extended, and now becomes Section 9. Its place as Section 4 is taken by a fuller treatment of polyanions, formerly briefly dealt with in a sub-section of Section 7. The rules for the nomenclature of inorganic boron compounds are outlined in Section 11. Extended tentative rules will be found in the *IUPAC Information Bulletin*: Appendices on Tentative Nomenclature, Symbols, Units and Standards, No. 8 (September 1970). The alphabetical principle, already mentioned in connection with Section 7, has also now been widely adopted in Sections 2 and 6.

The introduction of a preamble will, it is hoped, make clear more readily than is possible within the context of the formal rules, the precise meaning of terms such as oxidation number and coordination number, and the conventions governing the use of multiplying affixes, enclosing marks, numbers and letters. The most important tables have been placed together at the end of the rules and numbered, and an index has been added.

Introduction to the 1990 Recommendations

This volume is not a revision of the second (1970) edition but a completely new version presented in a new way which it is hoped will be much more useful to the general reader. This Part expounds the basic principles of inorganic chemical nomenclature, and is a general expansion of the 1970 edition. Part II and subsequent volumes will deal with specialized areas of nomenclature, and some of the eventual contents have already appeared in *Pure and Applied Chemistry*.

Chapters 1 and 2 are innovations. Chapter 1 is a concise history of inorganic chemical nomenclature, designed for general interest. Chapter 2 is a summary of the usages of inorganic nomenclature. The Commission expects that this Chapter will be generally useful. However, readers are warned that it is best used in conjunction with any appropriate second Chapter.

It was decided to present this edition in an instructional format rather than as a series of numbered rules. It is our experience that users of nomenclature books generally operate by selecting examples which most nearly fit the case they are working on. To that end, we have provided numerous examples and attempted to make the running text more coherent and discursive than it might have been had it been a series of bald statements of rules.

Throughout the text we have used the 1–18 Group numbering of the Periodic Table which is currently preferred by both CNIC and IUPAC. The related controversy and the intentions of IUPAC are discussed in the Appendix to this volume.

At every stage we have attempted to maintain consistency, both internally and with other IUPAC recommendations, principally the *Nomenclature of Organic Chemistry*, 1979 edition. We are aware that there are some differences of practice, but these have usually been pointed out in the text. In the spirit of this desire for consistency, we have tried to avoid the use of the word radical in anything other than the context of free radicals. This cuts across well established lines of inorganic usage, concerning not only substituents, but common groups in inorganic structures (uranyl, phosphoryl, etc.) and also terms such as acid radical. Occasionally rather crude circumlocutions have had to be used to convey the desired meaning, but the intention is that the word radical shall be understood only in the sense of free radicals, and that substituent groups shall be recognized as different entities even when they carry the same name as the corresponding free radical.

Chapters 3, 4, 5, 7, 8, and 9 represent consolidation and amplification of the material in the 1970 edition. Chapter 6 is vastly expanded over the corresponding Chapter in the 1970 edition, consonant with the vast strides which have been made in solid state chemistry. However, it does not deal with all aspects of the subject, for which satisfactory nomenclatures still need to be established. Consideration of iso- and hetero-polyanions will be presented in Part II, although the material of that Chapter has already been published in *Pure and Applied Chemistry*. Chapter 10 is greatly expanded compared to the corresponding Chapter in the 1970 Red Book. It contains a definitive exposition of coordination nomenclature, which, being primarily an inorganic nomenclature, we have attempted to apply more generally than solely to coordination compounds, for example, to the nomenclature of oxoacids. It also contains some innovative material, not currently widely used but regarded by the Commission as required even for relatively simple nomenclature purposes. One example of this is the discussion of stereochemistry and stereochemical descriptors. Chapter 11 also represents a considerable expansion over the 1970 version. The area of boron nomenclature is contentious and, in the terms we have used, specialized, suggesting that this material might well be in Part II. We have presented a survey of the nomenclature of simpler boron systems because boranes are much discussed even at relatively elementary levels of chemistry, and because we wish to codify those basic principles which seem firmly established.

Readers may note some inconsistencies in the presentation of organic formulae in this edition. This is deliberate. Representations such as Me, Et, and Ph are allowed without any obligation to define them, and they are used on occasion in this text in place of CH_3, C_2H_5, and C_6H_5. In addition, organic names are sometimes presented with locants as prefixes, and at other times the locants are infixes. The Nomenclature of Organic Chemistry, 1979 edition, specifically refrains from a firm ruling on this matter, and though it may ultimately be decided that locants should appear as prefixes we have decided not to be entirely rigorous about this matter. To some extent, therefore, the text reflects the predilections of individual authors.

In the second edition, the Commission used the terms 'Stock number' and 'Ewens-Bassett number' throughout the text. It has been decided that the terms 'oxidation number' and 'charge number' are preferable, being more immediately comprehensible, and references to the Stock and Ewens-Bassett numbers are made only parenthetically and on occasion for purposes of clarification. It is hoped that these terms will be replaced entirely by oxidation and charge number.

Readers should note that Chapters of which the designation begins with a roman I (e.g., Chapter I-3) belong to this volume, Part I. Chapters beginning with the designation II will appear in Part II. Tables referred to throughout this book receive simple sequential roman numerals. These Tables are collected together at the end of the volume. Those relating to specific Chapters, and the Figures, are numbered so that the Chapter to which they pertain is specified (e.g., Table I-6.2 is the second Table of Chapter 6).

Finally, care should be taken when considering names which begin on one line and spread onto a second. Such names naturally require a hyphen, the normal symbol indicating connectivity between separated parts of words. Because the hyphen is also a nomenclature device, confusion may arise. Although every care has been taken to minimize any such problems, readers should assure themselves of which hyphens are to be retained and which omitted when a split name is to be reproduced on one line.

Nomenclature of Inorganic Chemistry
RECOMMENDATIONS 1990

I-1   General Aims, Functions, and Methods of Chemical Nomenclature

CONTENTS

I-1.1   HISTORY AND AIMS OF CHEMICAL NOMENCLATURE

The activities of alchemy and of the technical arts practised prior to the founding of a true science of chemistry produced a rich vocabulary for describing chemical substances, but the names for individual species gave little indication of composition. Almost at the outset of the establishment of a true science of chemistry, in 1782, a 'system' of chemical nomenclature was developed by Guyton de Morveau (Note 1a). Guyton's statement of the need for a 'constant method of denomination, which helps the intelligence and relieves the memory' clearly defines the aims of chemical nomenclature. His system was extended by the joint contribution (Note 1b) of himself, Lavoisier, Berthollet, and Fourcroy and was popularized by Lavoisier (Note 1c). Later, Berzelius championed Lavoisier's ideas and adapted the nomenclature to the germanic languages (Note 1d). He expanded the system and added many new terms. This system was formulated before the enunciation of the atomic theory by Dalton and was based upon the concept of elements forming compounds with oxygen; the oxides in turn react with each other to form salts. The two-word names in some ways resembled the binary system introduced by Linnaeus (Carl von Linné) for plant and animal species.

Note 1a. Guyton de Morveau, L. B., *J. Phys.*, **19**, 310 (1782); *Ann. Chim. Phys.* **1**, 24 (1798).
Note 1b. Guyton de Morveau, L. B., Lavoisier, A. L., Berthollet, C. L., and de Fourcroy, A. F., *Méthode de Nomenclature Chimique*, Paris, 1787.
Note 1c. Lavoisier, A. L., *Traité Elémentaire de Chimie*, Third Edn, Deterville, Paris, 1801, Vol. I, pp. 70–81, and Vol. II.
Note 1d. Berzelius, J. J., *J. Phys.*, **73**, 248 (1811).

When atomic theory developed to the point where it was possible to write specific formulae for the various oxides and other binary compounds, names reflecting composition more or less accurately then became common. However, no names reflecting the composition of the oxosalts were ever adopted. Although the number of inorganic compounds grew considerably during the nineteenth century, the essential pattern of nomenclature was little altered until near the end of that century. As a need arose, a name was proposed, and nomenclature grew by accretion rather than by systematization.

When Arrhenius focused attention on ions as well as on molecules, it became necessary to name charged particles in addition to neutral species. It was not deemed necessary to develop a new nomenclature for salts; cations were designated by the names of the appropriate metal and anions by a modified name of the non-metal portion.

Along with the theory of coordination, Werner (Note 1e) proposed a system of nomenclature for coordination compounds which not only reproduced the compositions of compounds but also indicated the structures of many. Werner's system for coordination compounds was completely additive in that the names of the ligands were cited, followed by the name of the central atom modified by the suffix '-ate' if the complex was an anion. Werner also used structural descriptors and locants. The coordination nomenclature system was capable of expansion and adaptation to new compounds and even to other fields of chemistry.

The 1892 Geneva conference laid the basis for an internationally accepted system of organic nomenclature, but there was nothing comparable in the history of inorganic nomenclature. Probably for this reason, many *ad hoc* systems were developed for particular purposes but were not devised to be general. Thus, there often developed two or more methods for naming a given compound belonging to a given class. Each name may have value in a specific situation, or be preferred by some users. Unfortunately, this may also lead to confusion.

The primary aim of chemical nomenclature is simply to provide methodology for assigning descriptors (names and formulae) to chemical substances so that they can be identified without ambiguity, and thereby to facilitate communication.

A subsidiary aim is the achievement of standardization. This need not be so absolute as to require only one name for a substance; but the various 'acceptable' names should be kept to a manageable number.

It is recognized that the public needs and usage of nomenclature should be borne in mind. In some cases, the only requirement is to identify a substance; this was essentially the requirement of chemical nomenclature prior to the development of nomenclature 'systems' in the late 18th century. Even today, trivial names (Note 1f), abbreviations, codes, and the like are used by small groups of specialists. Within these groups, this may suffice as long as the members understand the device used for identification. However, this is not nomenclature as defined above, since such names do not necessarily convey structural and compositional information to a very wide audience by a recognisable, unambiguous, and general technique. To be widely useful, nomenclature must fulfill these requirements, and the unnecessary use of common names ('trivial' nomenclature), codes, and abbreviations in the formal scientific language is discouraged.

Note 1e. Werner, A., *Neuere Anschauungen auf den Gebieten der Anorganischen Chemie*, Third Edn, Vieweg, Braunschweig, 1913, pp. 92–95.

Note 1f. The adjective 'trivial' is used in the nomenclature sense of non-systematic, and is in no way perjorative.

I-1.2   FUNCTIONS OF CHEMICAL NOMENCLATURE

The first level of nomenclature beyond assignment of totally trivial names is names which give some systematic information about a substance but do not allow the inference of composition. Most of the common names of the oxoacids (e.g., sulfuric acid, perchloric acid) and of their salts are of this type. Such names are qualified as 'semi-systematic'. As long as they are for commonly used materials and understood by chemists, such names can be accepted, although they may hinder compositional understanding by those who have limited chemical training.

When a name itself allows the inference of the stoichiometric formula of a compound according to general rules, it becomes truly 'systematic'. Only a name at this second level of nomenclature becomes suitable for retrieval purposes.

Chemists today often concern themselves with the three-dimensional structures of substances, and the desire to incorporate this type of information into the name has grown rapidly. The systematization of nomenclature has had to expand to a third level of sophistication in order to accomplish this. Few chemists want to use such a degree of sophistication every time they refer to a compound, but they may wish to do so when appropriate.

A fourth level of nomenclature may be required for those who must compile and use extensive indexes. The cost to both compiler and searcher of multiple entries for a given substance is prohibitive. Accordingly, it is necessary in this context to develop systematic hierarchical rules that yield a unique name for a given substance.

A single compound will therefore have a multiplicity of names, depending upon the degree of specification required.

I-1.3   METHODS OF INORGANIC NOMENCLATURE

I-1.3.1   Formulation of rules

The Commission on the Nomenclature of Inorganic Chemistry (CNIC) studies all aspects of nomenclature of inorganic substances, making recommendations as to the most desirable practices, systematizing trivial methods, and proposing desirable practices to meet specific problems. These recommendations and proposals are for writing formulae and generating names for inorganic substances.

The Commission believes that nomenclature rules should be formulated precisely, to provide a systematic basis for assigning names and formulae within the defined area of the application of the rule. In doing so, extension of the rules to cover existing and emerging chemistry must be taken into account. New rules should be consistent as far as possible with existing recommended inorganic nomenclature and with nomenclature in other areas of chemistry. Recommendations incorporate common usage when it is systematic and unambiguous. Indeed, nomenclature developed in isolation from experimental chemistry will be perceived as an imposition and therefore irrelevant to science. At the same time, development of new rules may also require a more rigorous definition of existing rules to avoid inconsistencies, ambiguities, and proliferation of names. The revision of nomenclature is a continuous process as new discoveries make fresh demands on nomenclature systems.

I-1.3.2   **Name construction**

Systematic naming of an inorganic substance involves the construction of a name from units which are manipulated in accordance with defined procedures to provide composit- ional and structural information. The element names (or roots derived from them or from their Latin equivalents) are combined with appropriate affixes in order to construct systematic names by a variety of procedures which are called 'systems of nomenclature'.

There is a number of accepted systems for construction of names, as discussed in Section I-1.3.3. Perhaps the simplest system is that used for naming binary substances. The set of rules leads to a name such as iron dichloride for the substance $FeCl_2$; this name involves the juxtaposition of element names (iron, chlorine), their ordering in a specific way (electropositive before electronegative), the modification of an element name to indicate charge (the -ide ending designates an anion), and the use of the numerical prefix di- to indicate composition.

This and other patterns are commonly used in systematic inorganic nomenclature. Whatever is the pattern of nomenclature, names are constructed from units which fall into the following classes.

Element name roots
Numerical prefixes
Locants
Prefixes indicating atoms or groups – either substituents or ligands
Suffixes indicating charge
Suffixes indicating characteristic substituent groups
Infixes
Additive prefixes
Subtractive suffixes/prefixes
Descriptors (structural, geometric, stereochemical, etc.)
Punctuation

The uses of these units are summarized in Chapter I-2. The methods to assemble them differ, and the mastery of chemical nomenclature requires knowledge of these methods.

I-1.3.3   **Systems of nomenclature**

I-1.3.3.1   *General*

In the development of systematic nomenclature, several systems have emerged for the construction of chemical names. Each system has its inherent logic and its set of rules ('grammar') for elaborating chemical names. Some systems are more broadly applicable than others; however, practice has encouraged the use of specialized systems in particular areas of chemistry.

The existence of several distinct nomenclature systems has led to logically consistent alternative names for a given substance. This flexibility is highly useful in some contexts, but excessive proliferation of 'acceptable' alternatives can hamper communication, pose significant problems in searching the literature, and impede trade and legislation procedures. For this reason, the Commission endeavours to propose preferences.

In addition, confusion occurs when the 'grammar' of one nomenclature system is mistakenly used in another. This leads to profusion of apparently systematic names that are not really representative of any given system.

Three nomenclature systems are of primary importance to inorganic chemists.

I-1.3.3.2 *Binary-type nomenclature*

In this system, the composition of a substance is specified by the juxtaposition of element group names, modified or unmodified, together with appropriate numerical prefixes, if considered necessary. It is termed binary nomenclature because of the structure of the system, but its use is not restricted to the names of binary compounds.

Examples:
1. sodium chloride
2. silicon disulfide
3. magnesium chloride hydroxide

Grammatical rules are required to specify ordering of components, use of numerical prefixes, and the modification of some element names (e.g., silicide, chloride). The stoichiometric system and its extensions provide a basis for naming inorganic compounds systematically to indicate composition when structures are unknown or when structural information is not required. An easy extension is to the naming of compounds containing structurally defined sub-units (see Example 8).

Examples:
4. sodium cyanide
5. ammonium chloride
6. uranyl difluoride
7. sodium acetate
8. dipotassium tetraoxosulfate

I-1.3.3.3 *Coordination nomenclature*

This is an additive system for inorganic coordination compounds which treats a compound as a combination of a central atom with associated ligands (see Chapter I-10).

Examples:
1. triamminetrinitrocobalt, $[Co(NO_2)_3(NH_3)_3]$
2. sodium pentacyanonitrosylferrate(2−), $Na_2[Fe(CN)_5NO]$

It is also applicable in many other areas. Due to the variety of compounds belonging to this family, a large set of 'grammatical' rules has been elaborated. These provide for ligand names, citation of ligand names before central atom names, orders of citation in names and formulae, designation of charge on species and oxidation number, designation of stereochemistry, and designation of point of ligation in complicated ligands, etc. Extension of this system to polynuclear and cluster compounds is a major area of current nomenclature work.

I-1.3.3.4   *Substitutive nomenclature*

This is the system used extensively for organic compounds, but it has also been used to name many inorganic compounds. It is often based on the concept of a parent hydride modified by substitution of hydrogen atoms by groups (radicals) (see the Nomenclature of Organic Chemistry, 1979 edition).

This system has come into use for naming compounds formally derived from hydrides of certain elements in Groups 14, 15, 16, and 17 of the Periodic Table (see Chapter I-3). This is because these elements, as does carbon, form chains and rings of great complexity which can have many derivatives, and the system obviates the necessity of naming the hydrogen atoms of the parent hydride. Replacement of hydrogen atoms by characteristic groups leads to a method of naming a vast number of compounds.

Examples:
  1. bromobutane
  2. difluorosilane
  3. trichlorophosphane

Extensive rules are required to name parent compounds and substituents, to provide an order of citation of substituent names, and to specify the positions of attachment of substituents.

It is clear that the rules of each of these basic systems may provide unequivocal names for a given compound.

Example:
  4. silicon tetrachloride (binary-type nomenclature)
     tetrachlorosilicon (coordination nomenclature)
     tetrachlorosilane (substitutive nomenclature)

Coordination nomenclature is perhaps the most generally applicable in inorganic chemistry, but substitutive nomenclature may be applied in areas in which it is appropriate. Binary-type nomenclature is useful for specifying the composition of simple compounds.

I-1.3.3.5   *Supplementary systems*

In addition to these basic systems, several other systems are used to assign names in specific cases. The principal ones are mentioned below.

1. Oxoacids and oxoanions constitute an area of particular concern to inorganic chemists, and coordination nomenclature can be used to name them. However, current names are at best semi-systematic, and are still based on the original names used by Lavoisier and his associates. For every element which forms an oxoacid, there is an acid name with the suffix -ic and the names of the derived salts have the suffix -ate. Although the names of other oxoacids of the same element may follow a general pattern, names indicating stoichiometry have never come into general use. The recommendations on nomenclature of these compounds are in Chapter I-9.

The conventional names for condensed acids and their salts also rarely give indications of stoichiometry, and should ideally be replaced by an alternative systematic nomenclature.

Examples:

1. dichromic acid, $H_2Cr_2O_7$
2. sodium *cyclo*-triphosphate, $Na_3P_3O_9$
3. phosphododecamolybdic acid, $H_3PMo_{12}O_{40}$

2. Replacement nomenclature, including the oxa–aza type used in organic chemistry, is also used in inorganic chemistry.

Examples:

4. 1,4,7-triazacyclononane, $(NHCH_2CH_2)_3$
    (N replacing CH)
5. dicarba-*closo*-pentaborane(5), $B_3C_2H_5$
    (C replacing BH)
6. trisodium tetrathioarsenate, Na_3AsS_4
    (S replacing O)

3. Functional class nomenclature is a system developed in organic chemistry but sometimes used for purely inorganic compounds.

Examples:

7. phosphoric anhydride, P_4O_{10}
8. sulfuric diamide, $SO_2(NH_2)_2$

4. Additive nomenclature in inorganic chemistry is not restricted to coordination nomenclature and it may be combined with functional class nomenclature.

Examples:

9. ammonia—boron trifluoride (1/1), $H_3N \cdot BF_3$
10. triphenylphosphine oxide, $(C_6H_5)_3PO$

5. Subtractive nomenclature is often used in organic chemistry and has now been applied in inorganic chemistry, particularly to boron compounds.

Examples:

11. de-*N*-methylmorphine (removal of CH_2)
12. 6-deoxy-α-D-glucopyranose (removal of –O–)
13. 4,5-dicarba-9-debor-*closo*-nonaborate(2−), $B_6C_2H_8$
    (loss of BH)

I-1.3.4   **Nomenclature for indexes**

An efficient index requires that each chemical substance be assigned a unique name. The names recommended by CNIC are not necessarily unique. However, names selected for indexes should be consistent with IUPAC recommendations and comprehensible to index users.

I-1.4   INTERNATIONAL COOPERATION ON INORGANIC
             NOMENCLATURE

The need for a uniform practice among English-speaking chemists was recognized as early as 1886 and resulted in agreements on usage by the British and American chemical societies. In 1913, the Council of the International Association of Chemical Societies appointed a commission of inorganic and organic nomenclature, but World War I abruptly ended its activities. Work was resumed in 1921, when the International Union of Pure and Applied Chemistry, at its second conference, appointed commissions on the nomenclature of inorganic, organic, and biological chemistry. The first comprehensive report of the inorganic commission appeared in 1940 (Note 1g). This had a significant effect on the systematization of inorganic nomenclature and made many chemists aware of the necessity for developing a more fully systematic nomenclature. Among the significant features of this initial report were the adoption of the Stock system for indicating oxidation states (Note 1h), the establishment of orders for citing constituents of binary compounds in formulae and in names, the discouragement of the use of bicarbonate, etc. in the names of acid salts, and the development of uniform practices for naming addition compounds.

These IUPAC rules were revised and issued as a small book in 1959 (Note 1i). This book included chapters on crystalline phases of variable composition, non-stoichiometric compounds, and polymorphism. A treatment of coordination compounds was also included. The report gave positions for the placement of the mass number and atomic number around the element symbol and touched on the nomenclature of isotopically modified compounds. A second revision appeared in 1971 (Note 1j). This embodied a major extension of coordination nomenclature, a brief treatment of boron hydrides, a modest expansion of the tables of ion and radical names, a table of prefixes, and a table of element radical names. The Ewens–Bassett system (Note 1k) for designation of ionic charges was incorporated. The orders for citation of names and symbols were limited to two: an alphabetical sequence, and an element sequence based upon a continuous line drawn through an eighteen-column form of the Periodic Table.

Since 1971, the Commission has released documents on boron hydrides and related compounds (Note 1l), hydrides of nitrogen and derived ions and ligands (Note 1m), systematic names for the heavy elements (Note 1n), isotopically modified compounds

Note 1g. Jorisson, W. P., Bassett, H., Damens, A., Fichter, F., and Remy, H., *Ber. Dtsch. Chem. Ges. A*, **73**, 53 (1940); *J. Chem. Soc.*, 1404 (1940); *J. Am. Chem. Soc.*, **63**, 889 (1941).
Note 1h. Stock, A., *Angew. Chem.*, **32**, 273 (1919); **33**, 78 (1920).
Note 1i. *Nomenclature of Inorganic Chemistry*, 1957 Report of CNIC, IUPAC, Butterworths, London, 1959; *J. Am. Chem. Soc.*, **82**, 5523 (1960).
Note 1j. *Nomenclature of Inorganic Chemistry*, Second Edn, Butterworths, London, 1971.
Note 1k. Ewens, R. V. G. and Bassett, H., *Chem. Ind.* (*London*), 131 (1949).
Note 1l. Nomenclature of Inorganic Boron Compounds, *Pure Appl. Chem.*, **30**, 683 (1972).
Note 1m. Nomenclature of Hydrides of Nitrogen and Derived Cations, Anions, and Ligands, *Pure Appl. Chem.*, **54**, 2545 (1982).
Note 1n. Naming of Elements of Atomic Numbers Greater than 100, *Pure Appl. Chem.*, **51**, 381 (1979).

Note 1o), and nomenclature for regular single-strand and quasi-single-strand inorganic and coordination polymers (Note 1p).

The Commission published a supplement to the 1970 rules (Notes 1q) entitled 'How to Name an Inorganic Substance', which included a greatly expanded table of names for ions, radicals, and ligands. The Commission, together with the Commission on the Nomenclature of Organic Chemistry, also published a provisional treatment of the nomenclature of organometallic compounds (Note 1r), including chains and rings with regular patterns of heteroatoms; compounds containing phosphorus, arsenic, antimony, and bismuth; organosilicon compounds; and organoboron compounds.

I-1.5   NOMENCLATURE RECOMMENDATIONS IN OTHER AREAS OF CHEMISTRY

Inorganic chemical nomenclature develops in concert with those of other fields of chemistry. In interdisciplinary areas, knowledge of nomenclature practices in organic, biological, and macromolecular chemistry is necessary. As a guide to inorganic chemists working in such areas, some additional references (Notes 1s, 1t, and 1u) to nomenclature areas abutting inorganic chemistry are included in the references to this Chapter.

Note 1o. Isotopically Modified Compounds, *Pure Appl. Chem.*, **53**, 1887 (1981).

Note 1p. Nomenclature of Regular Single-strand and Quasi Single-strand Inorganic and Coordination Polymers, *Pure Appl. Chem.*, **57**, 149 (1985).

Note 1q. *How to Name an Inorganic Substance.* A Guide to the Use of "*Nomenclature of Inorganic Chemistry*", Pergamon Press, Oxford, 1977.

Note 1r. *Nomenclature of Organic Chemistry*, Pergamon Press, Oxford, 1979, Section D.

Note 1s. *Compendium of Analytical Nomenclature, Definitive Rules*, 1987, Second Edn, Blackwell Scientific Publications, Oxford, 1987.

Note 1t. *Compendium of Chemical Terminology, IUPAC Recommendations*, Blackwell Scientific Publications, Oxford, 1987.

Note 1u. *Quantities, Units and Symbols in Physical Chemistry*, Blackwell Scientific Publications, Oxford, 1988.

I-2   Grammar

CONTENTS

I-2.1 INTRODUCTION

Chemical nomenclature may be considered to be a language. As such, it is made up of words and it should obey the rules of syntax.

In the language of chemical nomenclature, the simple names of atoms are the 'words'. As words are assembled to form a sentence, names of atoms are assembled to form names of chemical compounds. Syntax is the set of grammatical rules for building sentences out of words. In nomenclature, this syntax includes the use of symbols, such as dots, commas, and hyphens, the use of numbers for appropriate reasons in given places, and the order of citation of various words, syllables, and symbols.

Generally, nomenclature systems use a base on which the name is constructed. This base can be derived from a parent compound name such as *sil* (from silane) in substitutive nomenclature (mainly used for organic compounds) or from a central atom name such as *cobalt* in additive nomenclature (mainly used in coordination chemistry).

Names are constructed by joining other units to these base components. Among the most important units are affixes. These are syllables or numbers added to words or roots and they can be suffixes, prefixes, or infixes, according to whether they are placed after, before, or within a word or root. Representative examples are listed in Table IX, together with their meanings.

Suffixes (endings) are of many different kinds, each of which conveys specific information. The following examples illustrate particular uses. They may specify the degree of unsaturation of a parent compound in substitutive nomenclature: hex*ane*, hex*ene*, hex*yne*; and phosph*ane*, phosph*ene*, phosph*yne*. Other endings indicate the nature of the charge carried by the whole compound; cobalt*ate* refers to an anion. Further suffixes can indicate that a name refers to a radical or group, as in hex*yl*, cobalt*io*.

Prefixes indicate, for example, substituents in substitutive nomenclature, as in the name *chloro*trisilane, and ligands in additive nomenclature, as in the name *aqua*cobalt. The prefixes can also be numbers to express specific information such as a point of attachment, e.g., 2-chlorotrisilane; or they can be multiplicative prefixes (Table III) to indicate the number of constituents or ligands, e.g., *hexa*aquacobalt.

Geometrical prefixes may be placed before the name to describe the geometry of the species. These prefixes are listed in Table V. Other devices may be used to complete the description of the compound. These include the charge (Ewens-Bassett) number to indicate the ionic charge, e.g., hexaaquacobalt(2+), and, alternatively, the oxidation (Stock) number to indicate the oxidation state of the central atom, e.g., hexaaquacobalt(II).

The designation of central atom and ligands, generally straightforward in mononuclear complexes, is more difficult in polynuclear compounds where there are several central atoms in the compound to be named. Then a priority order, sequential order, seniority order, or hierarchy has to be established. This problem arises in polynuclear coordination compounds, polyoxoanions, and chain and ring compounds. A hierarchy of functional groups is an established feature of organic nomenclature. For example, Table IV shows one of the seniority sequences used in inorganic nomenclature.

The various devices used in names (or formulae) are described successively below, together with their meanings and fields of application.

The purpose of this Chapter is to guide the users of nomenclature in building the name or formula of an inorganic compound and to help them verify that the derived name or formula fully obeys the accepted principles.

I-2.2   ENCLOSING MARKS

I-2.2.1   General

Chemical nomenclature employs three types of enclosing mark: braces { }, square brackets [ ], and parentheses ( ). The last two are frequently used. Where it is necessary in an inorganic FORMULA to use several sets of enclosing marks, square brackets, parentheses, and braces are used in the following nesting order of enclosing marks, [( )], [{( )}], [{[( )]}], [{{[( )]}}], etc. When it is necessary in an inorganic NAME to use several sets of enclosing marks the nesting order is different, {{{[( )]}}}, etc.

I-2.2.2   Square brackets

I-2.2.2.1   *Use in formulae*

Square brackets are used in FORMULAE in the following ways.
(a) To enclose the whole complex entity of a neutral coordination compound.

Examples:
1. $[Fe(C_5H_5)_2]$
2. $[PtCl_2(C_2H_4)(NH_3)]$

No numerical subscript should follow the square bracket used in this context. For example, for the ethylene derivative of $PtCl_2$, where the molecular formula is double the empirical formula, this should be indicated inside the square bracket.

Example:
3. $[\{PtCl_2(C_2H_4)\}_2]$ is more informative than $[Pt_2Cl_4(C_2H_4)_2]$. The representation $[PtCl_2(C_2H_4)]_2$ is incorrect.

(b) To enclose a complex ion. In this case, the superscript showing the charge appears outside the square bracket as do any subscripts indicating the number of complex ions in the salt.

Examples:

4. $[Al(OH)(H_2O)_5]^{2+}$
5. $Ca[AgF_4]_2$

(c) To enclose isopolyanions and heteropolyanions.

Examples:

6. $[S_2O_5]^{2-}$
7. $[PW_{12}O_{40}]^{3-}$

(d) To enclose structural formulae.

Examples:

8. $[O_2HPOPHO_2]^{2-}$

9.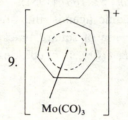

(e) To enclose a complex ion in a salt formula.

Examples:

10. $K[PtCl_3(C_2H_4)]$
11. $[Co(NH_3)_6] [Cr(CN)_6]$
12. $[Co\{SC(NHMe)_2\}_4](SO_4)$
13. $[Co(N_3)(NH_3)_5]SO_4$
14. $[CoCl_2(en)_2](NO_3)$ (en = ethane-1,2-diamine)

(f) In solid-state chemistry square brackets indicate an atom or a group of atoms in an octahedral site.

Example:

15. $(Mg)[Cr_2]O_4$

(g) In specifically labelled compounds.

Example:

16. $H_2[^{15}N]NH_2$

Note that this distinguishes the labelled compound from the substituted compound, $H_2{}^{15}NNH_2$.

(h) In selectively labelled compounds.

Example:

17. $[^{18}O, {}^{32}P]H_3PO_4$

I-2.2.2.2 *Use in names*

Square brackets are used in NAMES in the following ways.

(a) In the names of specifically and selectively labelled compounds the nuclide symbol is placed in square brackets before the name of the part of the compound that is isotopically modified. For isotopically substituted compounds, however, refer to Section I-2.2.3.2(h).

Examples:

1. $[^{15}N]H_2[^2H]$ $[^2H_1,{}^{15}N]$ammonia

2. $[^{13}C][Fe(CO)_5]$ $[^{13}C]$pentacarbonyliron

For more details, see Section I-4.7.2 and Nomenclature of Inorganic Chemistry, Isotopically Modified Compounds, *Pure Appl. Chem.*, **53**, 1887 (1981).

(b) Organic nomenclature is used for naming organic ligands and organic parts of coordination compounds. For such purposes, the use of square brackets obeys the principles stated in the *Nomenclature of Organic Chemistry*, 1979 edition.

I-2.2.3 **Parentheses**

I-2.2.3.1 *Use in formulae*

Parentheses are used in FORMULAE as follows.

(a) To enclose sets of identical groups of atoms (the entity may be an ion, radical, or molecule). Usually a multiplicative index follows the closing parenthesis. In the case of simple oxoions, parentheses are not mandatory.

Examples:

1. $Ca_3(PO_4)_2$

2. $B_3H_3(NCH_3)_3$

3. $[Ni(CO)_4]$

4. $(HBO_2)_n$

5. $(NO_3)^-$ or NO_3^-

6. $[Fe(H_2)H(Ph_2PCH_2CH_2PPh_2)_2]^+$

(b) To enclose the formula of a moiety which is an atom or set of atoms forming a neutral or a charged ligand in a coordination compound. The purpose is to separate the ligands from each other or from the remaining part of the molecule in order to avoid ambiguity. Parentheses may be used even if a multiplicative suffix is not needed.

Example:

7. $[Co(ONO)(NH_3)_5]SO_4$

(c) To enclose the abbreviation of a ligand name in compositional formulae. A list of recommended ligand abbreviations is included in Table I-10.5 and Table X.

Example:

8. $[\text{Co(en)}_3]^{3+}$

(d) In formulae in solid-state chemistry, to enclose symbols of atoms occupying the same type of site in a random fashion. The symbols themselves are separated by a comma, with no space.

Example:

9. K(Br,Cl)

(e) In solid-state chemistry, to indicate an atom or a group of atoms in tetrahedral sites.

Example:

10. $(\text{Mg})[\text{Cr}_2]\text{O}_4$

(f) To indicate the composition of a non-stoichiometric compound.

Examples:

11. $\text{Fe}_{3x}\text{Li}_{4-x}\text{Ti}_{2(1-x)}\text{O}_6$ $\quad(x=0.35)$
12. LaNi_5H_x $\quad(0<x<6.7)$

(g) In the Kröger–Vink notation (see Chapter I-6), to indicate a complex defect.

Example:

13. $(\text{Cr}_{\text{Mg}}V_{\text{Mg}}\text{Cr}_{\text{Mg}})^x$

(h) For crystalline substances, to indicate the type of crystal formed.

Examples:

14. $\text{ZnS}(c)$ $\quad$ *(c)* stands for cubic
15. $\text{AuCd}(CsCl$ type)

(i) In optically active compounds to enclose the signs of rotation, or the symbols for absolute configuration.

Example:

16. $(+)_{589}[\text{Co(en)}_3]\text{Cl}_3$

I-2.2.3.2 *Use in names*

Parentheses are used in NAMES in the following ways.
(a) Following multiplicative prefixes such as bis and tris, unless a nesting order is to be used (see Section I-2.2.1).

Example:

1. $[\text{CuCl}_2(\text{CH}_3\text{NH}_2)_2]$ $\quad$ dichlorobis(methylamine)copper(II)

(b) To enclose oxidation (Stock) and charge (Ewens-Bassett) numbers.

Example:

2. $Na[B(NO_3)_4]$ sodium tetranitratoborate(III), or
sodium tetranitratoborate(1−)

(c) In names for addition compounds and clathrates, to enclose stoichiometric ratios.

Example:

3. $8H_2S \cdot 46H_2O$ hydrogen sulfide—water (8/46)

(d) To enclose italic letters representing bonds between two (or more) metal atoms in coordination compounds.

Example:

4. $[Os_3(CO)_{12}]$ *cyclo*-tris(tetracarbonylosmium)(3 *Os—Os*)

(e) To enclose the stereochemical descriptors (see Chapter I-10)

Example:

5. $[Co(NO_2)_3(NH_3)_3]$ (*OC*-6-22)-triamminetrinitrocobalt(III)

(f) To enclose inorganic ligand names such as (triphosphato) which contain numerical prefixes, and for thio-, seleno-, and telluro-analogues of oxoanions which contain more than one atom, such as (thiosulfato), in order to avoid ambiguity. (Thio)(sulfato) means two different ligands.
(g) To enclose all organic ligand names whether they are neutral or not, or whether they are substituted or not, e.g., (benzaldehyde), (benzoato), when necessary, in order to avoid ambiguity. Where the ligands names themselves contain parentheses, it may be necessary to use a higher order of enclosing marks.

Example:

6. $[PtCl_3(C_2H_4)]^-$ trichloro(ethylene)platinate(II) ion

This formulation avoids confusion with the possibility that trichloroethylene is a ligand on platinum.
(h) In isotopically substituted compounds, to indicate the appropriate nuclide symbol(s) before the name for that part of the compound that is isotopically substituted [see *Pure Appl. Chem.*, **53**, 1887 (1981)]. Compare with the use of square brackets described in Section I-2.2.2.2(a).

Example:

7. H^3HO ($^3H_1$)water

(i) To enclose the number of hydrogen atoms in boron compounds.

Example:

8. B_6H_{10} hexaborane(10)

I-2.2.4 **Braces**

These are used in NAMES and FORMULAE within the hierarchical sequence outlined in Section I-2.2.1.

I-2.3 **HYPHENS, MINUS AND PLUS SIGNS, LONG DASHES, AND BOND INDICATORS**

I-2.3.1 **Hyphens** (Note 2a)

Hyphens are used in FORMULAE and in NAMES. Note that there is no space on either side of a hyphen.

(a) To separate symbols such as μ(mu), η(eta), and κ(kappa) from the rest of the formula or name.

(b) To separate from the rest of the formula or name geometrical or structural and stereochemical designators such as *cyclo, catena, triangulo, quadro, tetrahedro, octahedro, closo, nido, arachno, cis* and *trans,* and *(OC-*6-42), (Δ), and (λ). In dealing with aggregates or clusters, locant designators are similarly separated.

Example:

1.
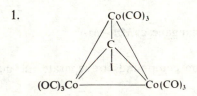

μ_3-iodomethylidyne-*cyclo*-tris(tricarbonylcobalt)(3 *Co—Co*)

(c) To separate locant designators from the rest of the name.

Example:
2. $Si_3H_5Cl_3$ 1,2,3-trichlorotrisilane

(d) To separate the labelling nuclide symbol from its locant in a selectively labelled compound.

Example:
3. $[1\text{-}^2H_{1;2}]SiH_3OSiH_2OSiH_3$

(e) To separate the name of a bridging ligand from the rest of the name.

Example:
4. $[Fe_2(CO)_9]$ tri-μ-carbonyl-bis(tricarbonyliron)

Note 2a. Often no distinction is made between minus sign and hyphens in texts. However, distinction is observed here. The minus sign is longer than the hyphen.

I-2.3.2   **Plus and minus signs**

The signs $+$ and $-$ are used to indicate the charge on an ion in a formula or name.

Examples:
1. Cl^-
2. Fe^{3+}
3. SO_4^{2-}
4. tetracarbonylcobaltate$(1-)$

They can also indicate the sign of optical rotation in the formula or name of an optically active compound.

Example:
5. $(+)_{589}[Co(en)_3]^{3+}$ $(+)_{589}$-tris(ethane-1,2-diamine)cobalt$(3+)$

I-2.3.3   **Long dashes**

(a) Long dashes (length two M) are used in FORMULAE in inorganic nomenclature only when the formulae are structural. In NAMES, similar dashes are used when it is desired to indicate metal–metal bonds in polynuclear compounds. They separate the italicized symbols of the bond partners which are contained in parentheses at the end of the name.

Example:
1. $[Mn_2(CO)_{10}]$ bis(pentacarbonylmanganese)$(Mn—Mn)$

(b) They are used in names of addition compounds to separate the molecular constituents.

Examples:
2. $3CdSO_4 \cdot 8H_2O$ cadmium sulfate—water (3/8)
3. $2CHCl_3 \cdot 4H_2S \cdot 9H_2O$ chloroform—hydrogen sulfide—water (2/4/9)

I-2.3.4   **Special bond indicators for line formulae**

This structural symbol ⌐‾‾‾¬ or ∟___⌐ (a long square bracket facing downwards or upwards) may be used in line formulae when necessary to indicate bonds between non-adjacent atom symbols.

Examples:
1. $[\overline{NiS}=P(CH_3)_2(\eta^5\text{-}C_5H_5)]$

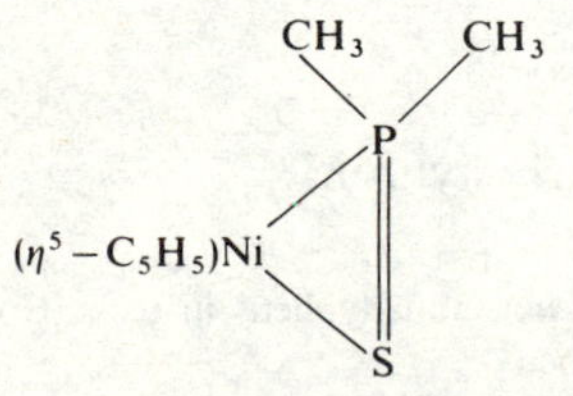

2. $[(CO)_4\overline{MnMo(CO)_3\{(\eta^5\text{-}C_5H_4)}P(C_6H_5)_2\}]$

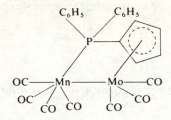

I-2.4   SOLIDUS

The solidus (/) is used in names of addition compounds to separate the arabic numerals which indicate the proportions of individual constituents in the compound.

Examples:
1. $BF_3 \cdot 2H_2O$ boron trifluoride—water (1/2)
2. $BiCl_3 \cdot 3PCl_5$ bismuth trichloride—phosphorus pentachloride (1/3)

I-2.5   DOTS, COLONS, COMMAS, AND SEMICOLONS

I-2.5.1   Dots

Dots are used in FORMULAE in various positions.
(a) As right superscripts they indicate unpaired electrons in radicals (see Section I-4.4.3). For transition metal complexes such indication is often deemed unnecessary.

Example:
1. $[V(CO)_6]^{\bullet}$

(b) As right superscripts in the Kröger–Vink notation of solid state chemistry, they indicate the unit of positive effective charge (one dot for every charge unit).

Example:
2. $Li^x_{Li,\,1-2x}Mg^{\bullet}_{Li,\,x}V'_{Li,\,x}Cl^x_{Cl}$

(c) Centre dots in FORMULAE of hydrates, addition compounds, double salts, and double oxides separate the individual constituents. The dot is written in the centre of the line to distinguish it from a full stop (period).

Examples:
3. $ZrCl_2O \cdot 8H_2O$
4. $NH_3 \cdot BF_3$
5. $CuCl_2 \cdot 3Cu(OH)_2$
6. $Ta_2O_5 \cdot 4WO_3$

I-2.5.2   **Colons**

(a) Colons are used in NAMES of coordination compounds to separate the ligating atoms of a ligand which bridges central atoms.

Example:

1. $[\{Co(NH_3)_3\}_2(\mu\text{-}NO_2)(\mu\text{-}OH)_2]Br_3$

   di-μ-hydroxo-(μ-nitrito-κN:κO)bis(triamminecobalt)(3+)   tribromide

(b) In names of boron compounds to separate the sets of locants of boron atoms which are connected by bridging hydrogen atoms (see Chapter I-11).

Example:

2. B_9H_{15} (3,4:3,9:5,6:6,7:7,8-penta-μH)-(3-*endo-H*)-*nido*-nonaborane

(c) To separate the italicized element symbols in the constitutional repeating unit in an inorganic polymer. See Nomenclature for Regular Single-strand and Quasi Single-strand Inorganic and Coordination Polymers, *Pure Appl. Chem.*, **57**, 149 (1985).

I-2.5.3   **Commas**

Commas are used in NAMES in five ways.

(a) To separate locants: see Example 1 under Section I-2.5.4

(b) To separate the symbols of the ligating atoms of a polydentate chelating ligand.

Example:

1. *cis*-bis(glycinato-*N,O*)platinum

(c) In solid state chemistry, to separate symbols of atoms substituting each other.

Example:

2. $(Mo,W)_nO_{3n-1}$

(d) To separate oxidation (Stock) numbers in a mixed valence compound.

Example:

3. $[(NH_3)_5Ru(C_4H_4N_2)Ru(NH_3)_5]^{5+}$

   μ-pyrazine-bis(pentaammineruthenium)(II,III)

(e) In selectively labelled compounds to separate symbols of labelled atoms.

Example:

4. $[^{18}O, ^{32}P]H_3PO_4$ $[^{18}O,^{32}P]$phosphoric acid

I-2.5.4   **Semicolons**

Semicolons are used in at least three ways.

(a) In the names of coordination compounds, to order locants separated by commas, as in the kappa convention.

Example:

(1)   (2)
1. $[Cu(2,2'-bpy)(H_2O)(\mu-OH)_2Cu(2,2'-bpy)(SO_4)]$
    aqua-$1\kappa O$-bis(2,2'-bipyridine)-$1\kappa^2 N^1,N^{1'};2\kappa^2 N^1,N^{1'}$-di-$\mu$-hydroxo-
    [sulfato($2-$)-$2\kappa O$]dicopper(II)

(b) To separate the subscripts in order to indicate the number of possible positions in selectively labelled compounds.

Example:
2. $[1-{}^2H_{1;2}]\,SiH_3OSiH_2OSiH_3$

(c) In boron nomenclature, to separate sets of locants designating a hydrogen atom bridge (see Example 2, Section I-2.5.2).

I-2.6   SPACES

Spaces are used in names only according to defined rules; the practice may be different in languages other than English.

In NAMES they are used as follows.

(a) To separate the names of ions from each other in salt names.

Examples:
1. NaCl   sodium chloride
2. $NaTl(NO_3)_2$ sodium thallium(I) dinitrate

(b) In binary compounds, to separate the electropositive part of the name from the electronegative part.

Example:
3. P_2O_5 diphosphorus pentaoxide

(c) To separate the arabic numeral from the symbols of central atoms written in italics between parentheses at the end of the name of a polynuclear compound.

Example:
4. $[Os_3(CO)_{12}]$ *cyclo*-tris(tetracarbonylosmium)(3 *Os—Os*)

(d) In addition compounds to separate constituent proportions from the remainder of the name.

Example:
5. $3CdSO_4\cdot 8H_2O$ cadmium sulfate—water (3/8)

I-2.7 ELISIONS

In general, in inorganic nomenclature no elisions are made when using numerical prefixes, except for compelling linguistic reasons (Note 2b).

Example:
1. 'tetraaqua' and not 'tetraqua'

I-2.8 NUMERALS

I-2.8.1 **Arabic numerals**

Arabic numerals are the keys of nomenclature. Their place in a formula or in a name therefore has a very special significance. They are used in FORMULAE in many ways.
(a) As right subscripts to indicate the number of individual constituents (atoms or groups of atoms). Usually, unity is not indicated.

Examples:
1. $CaCl_2$
2. $[Co(NH_3)_6]Cl_3$

(b) As a right superscript to indicate the charge number.

Examples:
3. Cu^{2+}
4. $[Al(H_2O)_6]^{3+}$

(c) To indicate the composition of addition or non-stoichiometric compounds. The numeral is written on the line before the molecular formula of each constituent except that unity is omitted.

Examples:
5. $Na_2CO_3 \cdot 10H_2O$
6. $8WO_3 \cdot 9Nb_2O_5$

(d) To designate the mass number and/or the atomic number of nuclides represented by their symbols. The mass number is written as a left superscript, and the atomic number as left subscript.

Examples:
7. $^{18}_{8}O$
8. $^{3}_{1}H$

(e) To indicate the hapticity (denticity) of a ligand (number of atoms in a given ligand which are directly connected to a central atom in a coordination compound). It is written as a right superscript to the symbols η and κ (see Chapter I-10).

Note 2b. The exception is monoxide, preferred to monooxide.

Example:
9. $[Fe(\eta^5\text{-}C_5H_5)_2]$

Arabic numerals are also used in NAMES in many ways, as summarized below.
(a) To indicate the number of metal–metal bonds in polynuclear coordination compounds.

Example:
10. $[\{Ni(\eta^5\text{-}C_5H_5)\}_3(CO)_2]$
    di-μ_3-carbonyl-*cyclo*-tris(cyclopentadienylnickel)(3 *Ni—Ni*)

(b) To indicate the charge number.

Example:
11. $[CoCl(NH_3)_5]^{2+}$ pentaamminechlorocobalt(2+) ion

(c) To indicate bridging multiplicity.

Example:
12. $[\{(PtI(CH_3)_3\}_4]$ tetra-μ_3-iodotetrakis(trimethylplatinum)

(d) Arabic numerals are used in the nomenclature of boron compounds (see Chapter I-11) to indicate the number of hydrogen atoms in the parent borane molecule. The arabic numeral is enclosed in parentheses immediately following the name.

Examples:
13. B_2H_6 diborane(6)
14. $B_{10}H_{14}$ decaborane(14)

(e) As a superscript to the italicized symbol of the atom of a polydentate ligand; an arabic numeral is a locant for specific donor atoms. This rule is often applied in an *ad hoc* fashion, especially where the rules of organic chemistry do not provide specific atom numbering for the atoms of interest.

Example:

15.   tartrato(3−)-O^1,O^2

(f) In addition compounds, the numerals indicate the numbers of molecules of the constituents.

Example:
16. $8H_2S \cdot 46H_2O$ hydrogen sulfide—water   (8/46)

(g) In polynuclear structures, arabic numerals are part of the CEP descriptor for central structural units (see Section I-10.8.3.3).
(h) As a superscript to indicate the non-standard bonding number in the λ convention (see Chapter I-7).

Example:
17. IH_5 λ^5-iodane

I-2.8.2   **Roman numerals**

Roman numerals are used in FORMULAE as right superscripts to designate the oxidation number.

Examples:
1. $[Co^{II}Co^{III}W_{12}O_{42}]^{7-}$
2. $[Mn^{VII}O_4]^-$
3. $(Fe^{II}Fe^{III}_2)O_4$

In NAMES they indicate the oxidation number of an atom, and are enclosed in parentheses immediately following the name of the atom being qualified.

Example:
4. $[Fe(H_2O)_6]^{2+}$ hexaaquairon(II) ion

I-2.9   ITALIC LETTERS

Italic letters (Note 2c) are used in NAMES as follows.
(a) For the geometrical and structural affixes such as *cis*, *cyclo*, *catena*, *triangulo*, and *nido* (see Table V).
(b) To designate symbols of metal atoms bonded to other metal atoms in polynuclear coordination compounds.

Example:
1. $[Os_3(CO)_{12}]$ dodecacarbonyltriosmium(3 *Os—Os*)

(c) In double oxides and hydroxides when the structural type is to be indicated.

Note 2c. In typed manuscripts, where italic symbols are often unavailable, it is customary to indicate italic words or letters by an underline.

24

Example:

2. $MgTiO_3$ magnesium titanium trioxide   (*ilmenite* type)

(d) In coordination compounds italic symbols designate the atom or atoms of a ligand (usually polydentate) to which the metal is bound, whether the kappa convention is used or not.

Example:

3.

cis-bis(glycinato-*N*,*O*)platinum

(e) In solid state chemistry, in Pearson and crystal system symbols (see Section I-6.5).
(f) In names of coordination compounds, to designate polyhedral symbols using capital letters (see Section I-10.5.2).

Example:

4. $[Co(NO_2)_3(NH_3)_3]$ (*OC*-6-22)-triamminetrinitrocobalt(III)

Italic letters are also used in the nomenclature of inorganic chemistry, and quite generally, to represent numbers, especially in formulae, where the numbers are undefined, e.g., $(HBO_2)_n$, Fe^{n+}, etc.

I-2.10   GREEK ALPHABET

The Greek alphabet is of frequent use in inorganic chemical nomenclature. Some uses are summarized as follows below.

Δ is used for absolute configuration. It is used as a structural descriptor to designate deltahedra.

δ is used for the absolute configuration of chelate rings and in solid state nomenclature to indicate small variations of composition.

η is used as a symbol to designate the hapticity of a ligand (see Chapter I-10).

κ is used as a ligating atom designator in the *kappa convention* (see Chapter I-10).

Λ is used for absolute configuration.

λ is a symbol, usually bearing an arabic numeral as a right superscript, used to indicate non-standard bonding number in the *lambda convention* [see *Pure Appl. Chem.*, **54**, 217 (1982)]. It also designates the helical character of the conformation of chelate rings.

μ is used as a symbol for bridging ligands.

Note that the use of σ and π to indicate bonding type, and as suggested in earlier versions of the *Nomenclature of Inorganic Chemistry*, is no longer recommended.

I-2.11   ASTERISKS

The asterisk (*) is used in FORMULAE as right superscript to the symbol of an element, in the following ways.

(a) It can specify a chiral centre.

Example:

1.

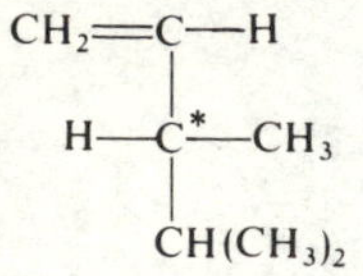

$$CH_2 = C - H$$
$$H - C^* - CH_3$$
$$CH(CH_3)_2$$

This usage has been extended to label a chiral ligand or a chiral centre in coordination chemistry.

Examples:
  2. L* = (+)-diop, or (4S,5S)-diop
     (for 'diop' see Table X)

3.

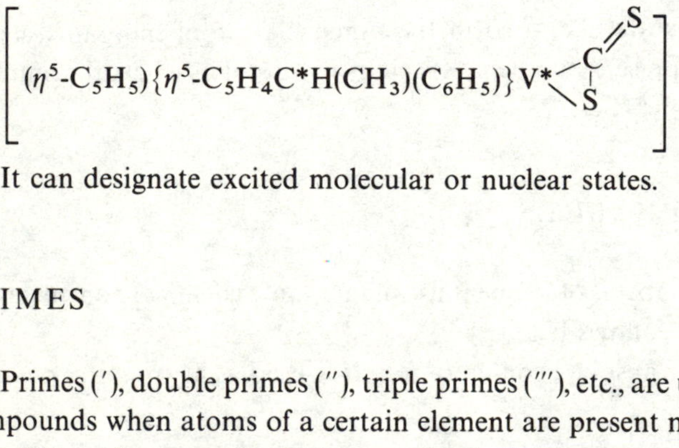

$$\left[(\eta^5\text{-}C_5H_5)\{\eta^5\text{-}C_5H_4C^*H(CH_3)(C_6H_5)\} V^* \underset{S}{\overset{S}{<}} C \right]$$

(b) It can designate excited molecular or nuclear states.

I-2.12   PRIMES

(a) Primes ('), double primes ("), triple primes ('''), etc., are used in names of coordination compounds when atoms of a certain element are present more than once in the organic ligand in such a manner that some or all of these atoms are ligated to the metal. The ligated atoms are primed successively in ascending order to distinguish them from the non-ligated atoms of this same element (however, see also Section I-10.6.2.1).

Example:
  1. $[Rh_3(CO)_3Cl(\mu\text{-}Cl)[\mu_3\text{-}\{(C_6H_5)_2PCH_2P(C_6H_5)CH_2P(C_6H_5)_2\}]]Cl$
     tricarbonyl-$1\kappa C,2\kappa C,3\kappa C$-$\mu$-chloro-$1:2\kappa^2 Cl$-chloro-
     $3\kappa Cl$-bis-μ_3-[bis[(diphenylphosphino)-$1\kappa P':3\kappa P''$-
     methyl]phenylphosphine-$2\kappa P$]-trirhodium(1+) chloride.

(b) Primes, double primes, triple primes, etc., are also used as right superscripts in the Kröger–Vink notation (see Section I-6.4), where they indicate a site which has one, two, or three units of negative effective charge, respectively.

26

Example:

2. $Li^x_{Li,\,1-2x}Mg^{\bullet}_{Li,\,x}V'_{Li,\,x}Cl^x_{Cl}$

I-2.13   MULTIPLICATIVE PREFIXES

The number of identical chemical entities in a name is expressed by a numerical prefix.

In the case of simple entities such as monoatomic ligands the multiplicative prefixes di-, tri-, tetra-, penta-, . . . are used. In the case of complex entities such as organic ligands (particularly if they are substituted) the multiplying prefixes bis-, tris-, tetrakis-, pentakis-, . . . are used, i.e., -kis is added starting from tetra-. The modified entity is often placed within parentheses to avoid ambiguity.

Examples:

1. $[PtCl_4]^{2-}$ tetrachloroplatinate(2−)ion
2. $[Fe(C_2C_6H_5)_2(CO)_4]$ tetracarbonylbis(phenylethynyl)iron

Composite numeral prefixes are built up by citing units first, then tens, then hundreds, and so on.

Example:

3. 35 is written pentatriaconta (or pentatriacontakis).

In the case of icosa, the letter i is elided in the cases of dicosa and tricosa. A list of prefixes is included in Table III. For detailed usage, see Chapters I-5, I-7 and I-10 [see also the extension of rules A-1.1 and A-2.5 concerning numerical terms used in organic chemical nomenclature, *Pure Appl. Chem.*, **55**, 1463 (1983)].

I-2.14   LOCANT DESIGNATORS

I-2.14.1   **Introduction**

In inorganic chemistry the molecular entity may have a skeletal shape which may be as simple as a square or as complex as a large polyhedron. In some cases less complex units are associated by sharing elements such as edges, vertices, and facial planes. Because of the variety of compounds in which association occurs, locant designators are used to assign a topological site to the central atoms in the array. These can be arabic numerals or lower case letters.

I-2.14.2   **Arabic numerals**

These are used in cases where no metallic element is present, as in boron or silicon compounds, and in chain and rings compounds. Organic nomenclature practice is followed to define the numbering order of the atoms of the molecule or ion in each family of compounds. This permits specification of the position in the skeletal backbone of the various groups attached to it. This is not the place to deal at length with all the cases where such locant designators occur. They are adequately described in the *Nomenclature of Organic Chemistry*, 1979 edition. The following example will, however, illustrate their use.

Example:

1. $[B_{12}H_{12}]^{2-}$

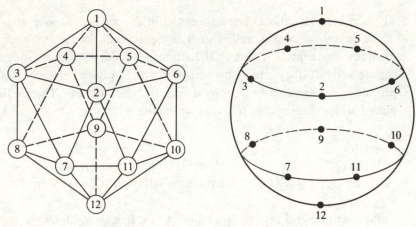

dodecahydro-*closo*-dodecaborate(2−)

(Boron atoms are located at the numbered positions; each boron atom carries one hydrogen atom).

Arabic numerals are also used as locants in polynuclear coordination compounds and in clusters (see Chapter I-10).

Example:

2. $[Co(CO)_4Re(CO)_5]$ nonacarbonyl-1$\kappa^5 C$,2$\kappa^4 C$-cobaltrhenium(*Co—Re*)

I-2.14.3   Lower case letters

These are used in polyoxoanion nomenclature. The central atoms of polyhedra (usually octahedra or tetrahedra) are numbered in the same way as boron compounds, but the vertices around central atoms need to be numbered as well. They are designated by a lower-case letter attached to the number of the central atom to which a particular vertex refers. An octahedron needs six letters a, b, c, d, e, f, for its six vertices, a tetrahedron needs four letters a, b, c, d, for its four vertices, etc. The resulting locant designation is shown on the following page for the case of $[Mo_6O_{19}]^{2-}$. A detailed treatment is given in Nomenclature of Polyoxoanions, *Pure Appl. Chem.*, **59**, 1529 (1987).

I-2.15   PRIORITIES (SENIORITIES)

I-2.15.1   Introduction

In chemical nomenclature the terms 'priority', 'seniority', and 'precedence' convey the notion of rank or order among a number of possibilities and are ubiquitous concepts of fundamental importance. Chemical nomenclature deals with elements and their combinations among themselves, either individually or as groups (element with element; group

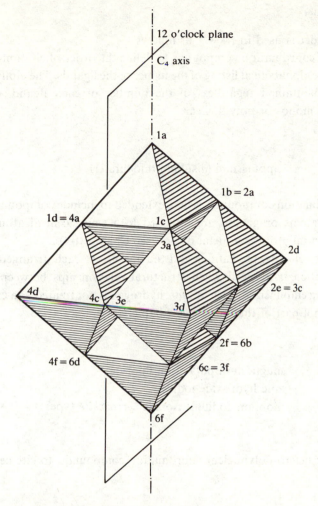

Locant designators in $[Mo_6O_{19}]^{2-}$ ion

with group). The groups of atoms can be ions, ligands in coordination compounds, or substituents in hydrides.

Whereas writing the symbol or the name of an element is straightforward, a choice of which element to write first in the FORMULA and in the NAME has to be made as soon as another element is to be associated with this element to form, for instance, a binary compound. The order of citation is based upon some established methods of choice which are outlined below (see Chapter I-4).

I-2.15.2   **Electronegativity criterion**

In formulae and names of binary compounds of nonmetallic elements, that constituent is cited first which appears earlier in the sequence shown below. Note that this is an approximate electronegativity sequence, though it departs in detail, for example, in the relative positions of C and H, from the sequence usually accepted.

Rn, Xe, Kr, Ar, Ne, He, B, Si, C, Sb, As, P, N, H, Te, Se, S, At, I, Br, Cl, O, F

Example:
1. S_2Cl_2

I-2.15.3   **Alphabetical order**

Alphabetical order is used in NAMES as follows.

(a) In names of coordination compounds to define the order of citation of ligands. This order follows the alphabetical listing of the names of the ligands. The alphabetical citation of ligands is maintained regardless of the number of each ligand, or whether the compounds are mono- or poly-nuclear.

Example:

1. $K[AuS(S_2)]$ potassium (*d*isulfido)*t*hioaurate(1−)

The term 'coordination compounds' is here extended to include compounds where two or more different atoms or groups are attached to a single central atom, regardless of whether this central atom is a metal or not (see Chapter I-4).

(b) In the names of salts, the cations and the anions are each arranged alphabetically. However, deviation is allowed when structural relationships between different compounds are being emphasized. In acid salts, hydrogen is not treated as a cation unless it is unequivocally not part of the anion (see Chapter I-8).

Examples:

2. $KMgF_3$ magnesium potassium fluoride
3. $ZnI(OH)$ zinc hydroxide iodide
4. $NaNbO_3$ niobium sodium trioxide (*perovskite* type)

(c) In names of hetero-polynuclear coordination compounds, to cite central atoms.

Example:

5. $[CoCu_2Sn(C_5H_5)(CH_3)\{\mu\text{-}(C_2H_3O_2)\}_2]$
   bis(μ-acetato)-cyclopentadienylmethylcobaltdicoppertin

Alphabetical order is used in FORMULAE as follows.

(a) In coordination compounds to arrange the sequence of ligands for each set of ionic and neutral ligands (ionic preceding neutral). Ligands within each set are cited in the alphabetical order of the first symbol of their formulae (see Sections I-4.6.7 and I-10.3.1). However, some deviation is allowed if it is desired to convey specific structural information.

Example:

6. $[CrCl_2(H_2O)_2(NH_3)_2]$

(b) In the formulae of polynuclear coordination compounds or polyanions, to cite central atoms of different chemical species.

Example:

7. $[Fe_2Mo_2S_4(C_6H_5S)_4]^{2-}$

(c) In the formulae of salts and double salts, to establish the sequential orders in cations and anions, respectively (see Chapter I-4).

Examples:
8. BiClO
9. $KNa_4Cl(SO_4)_2$

I-2.15.4 Element seniority sequence

This is a sequence based on the Periodic Table, which is widely available and used everywhere; it is shown in Table IV. The element columns (1 to 18) are connected by arrows leading in a direction starting from the 'less metallic' elements and moving towards the 'more metallic' elements. This order had its origin in electronegativity considerations. It is followed in the situations cited below.

(a) In numbering central atoms in polynuclear coordination compounds. The central atom *coming last* in the direction of the arrow is given the lowest number and the other atoms are numbered in the order they are encountered following the reverse direction of the arrow.

Example:

$$\qquad\qquad \overset{2}{} \qquad \overset{1}{}$$
1. $[Co(CO)_4Re(CO)_5]$ \qquad nonacarbonyl-$1\kappa^5C,2\kappa^4C$-cobaltrhenium(*Co—Re*)

(b) In substitutive nomenclature, where the names of compounds are derived from a parent hydride. Where different elements of Group 14 are present in the parent, as in H_3GeSiH_3, a choice of seniority is necessary. The choice is based on the element appearing later or last in Table IV (see the *Nomenclature of Organic Chemistry*, 1979 edition).

I-2.15.5 Other priority sequences

I-2.15.5.1 *Organic priority orders*

In organic nomenclature, an order for the choice of principal organic functional group named 'The class of characteristic groups' is used (see the *Nomenclature of Organic Chemistry*, 1979 edition, Rule C-10.1, etc.). When an organic group occurs in an inorganic compound, organic nomenclature is followed to name the organic part.

I-2.15.5.2 *Priority of ligand types*

In formulae of coordination compounds, when there are several types of ligand such as ionic and neutral, anionic ligands are cited before the neutral. Both CO and NO are considered neutral for nomenclature purposes. Bridging ligands are given last among the ligands, and are cited in increasing order of bridging multiplicity.

In names of coordination compounds, the names of the ligands precede that of the metal. Bridging ligands (listed in alphabetical order with the other ligands) are given before the corresponding non-bridging ligands, e.g., di-μ-chloro-tetrachloro, and multiple bridging is listed in descending order of complexity, e.g., μ_3-oxo-di-μ-trioxo . . . However, in inorganic and coordination polymers, bridging groups are cited at the end of the name, to comply with usual polymer practice [see *Pure Appl. Chem.*, **57**, 149 (1985)].

I-2.15.5.3   *Priorities in salt names and formulae*

In names and formulae of salts, double salts, and coordination compounds, cations precede anions. In NAMES of acid salts, hydrogen is not normally cited as a cation, since it is part of the anion name (see Sections I-4.6, I-5.3.2, I-8.5.2 and I-8.5.3).

I-2.15.5.4   *Isotope labelling and modification*

In isotopically modified compounds, a priority principle governs the order of citation of nuclide symbols [see *Pure Appl. Chem.*, **53**, 1887 (1981)].

I-2.15.5.5   *Stereochemical priorities*

In the stereochemical nomenclature of coordination compounds, the procedure for assigning priority numbers to the ligating atoms of a mononuclear coordination system is based upon the standard sequence rules developed for chiral carbon compounds (Cahn, Ingold, Prelog rules). For details see Chapter I-10.

I-2.15.5.6   *Priority sequences of punctuation marks*

In the names of coordination compounds and boron compounds, the punctuation marks used to separate the symbols of atoms from the numerical locants, the locants indicating bridging atoms, and the various other sets of locants which may be present, are arranged in the hierarchy shown below.

comma – junior to colon – junior to semicolon.

The colon is only used for bridging ligands, so that the more restricted general hierarchy is simply comma – junior to semicolon. The sequence when bridging ligands are being specified is comma – junior to colon (see Section I.2.5.2, Example 2, and Chapter I-11).

I-2.16   AFFIXES (Prefixes, suffixes, and infixes)

Any name more complex than a simple element name has a structure, a root with a prefix or a suffix. The suffix is a terminal vowel or a combination of letters. These terminations carry information and are very useful in shortening names and including some special meanings. Affixes of current usage are listed in Table IX and some prefixes are included in Table V.

Table IX is not exhaustive. Some terminations used either in organic chemistry or in biochemistry and seldom used in inorganic chemistry have been excluded. The first part of the table contains simple suffixes, i.e., those which carry a single item of information. The second part deals with combined affixes, i.e., combinations of several suffixes which, when used together at the end of a name, convey more than one piece of information.

I-2.17   FINAL REMARKS

This Chapter is intended to be a guide for the users of inorganic nomenclature. The gathering of the various usages in names and formulae under common headings provides the reader with an easy check to ensure that the constructed name or formula is in accordance with the agreed practice. However, it is not sufficient to make clear all the rules needed to build a name or a formula, and for this purpose the reader is advised to consult the appropriate Chapters for the detailed treatment.

I-3   Elements, Atoms, and Groups of Atoms

CONTENTS

I-3.1   INTRODUCTION

This Chapter is concerned with one of the basic formalisms of chemistry, the representation of the elements by symbols. Generally no distinction is made between an element and an elementary substance in English-speaking countries. However, some consider the first to be an abstraction whereas the second is undoubtedly a form of matter. Often it is not clearly indicated in usage whether a symbol stands for an atom or an element.

Considerable difficulty was experienced in obtaining definitions that satisfied all requirements. The definitions presented here are intended to be useful and widely applicable, even if they can sometimes be criticised from a philosophical point of view.

I-3.2   DEFINITIONS

I-3.2.1   **Element**

An ELEMENT (or elementary substance) is matter, all of whose atoms are alike in having the same positive charge on the nucleus (see Sections I-3.3 and I-3.6).

I-3.2.2   **Atom**

An ATOM is the smallest unit quantity of an element that is capable of existence whether alone or in chemical combination with other atoms of the same or other elements (see Section I-3.3 and Tables I and II).

I-3.2.3   **Atomic number**

The ATOMIC NUMBER of an atom is the number of electronic units of positive charge carried by the nucleus of that atom (see Tables I and II).

I-3.2.4   **Mass number**

The MASS NUMBER of an atomic nucleus is the total number of protons and neutrons in the nucleus (see Section I-3.4).

I-3.2.5   **Nuclide**

A NUCLIDE is any atomic species defined by specific values of atomic number and mass number.

I-3.2.6   **Isotopes**

ISOTOPES are two or more different nuclides having the same atomic number (see Section I-3.5).

I-3.2.7   **Allotropes**

ALLOTROPES of an element are different structural modifications of that element (see Section I-3.7).

I-3.2.8   Atomic symbol

The ATOMIC SYMBOL consists of one, two, or three letters used to represent the atom in chemical formulae (see Sections I-3.3.3, I-3.3.4, and I-3.3.5).

I-3.2.9   Groups of elements

GROUPS OF ELEMENTS are elements that have been grouped together on the basis of some similarity. Some groups bear collective names, e.g., alkali metals, halogens, etc. (see Section I-3.8).

I-3.3   NAMES AND SYMBOLS OF ATOMS

I-3.3.1   Introduction

The origins of the names of some chemical elements, such as antimony, are lost in antiquity. Other elements recognised (or discovered) during the past three centuries were named according to various arbitrary associations of origin, physical or chemical properties, etc., and more recently to commemorate the names of some famous scientists. In 1979 IUPAC approved a systematic nomenclature for elements of atomic numbers greater than one hundred (see Section I-3.3.5 and Table II).

In the course of time the names of the elementary substances became transferred to the corresponding atoms and these names provide the basis of the IUPAC-approved *Nomenclature of Inorganic Chemistry*.

I-3.3.2   Atoms of atomic number less than 101 and their names

The IUPAC-approved names of the atoms of atomic numbers 1–100 for use in the English language are listed together with those of atomic numbers 101–109 in alphabetical order in Table I. It is desirable that the names in different languages differ as little as possible. The names approved by IUPAC are based on considerations of practicality and prevailing usage. It is emphasised that the IUPAC selection carries no implication regarding priority of discovery.

Other names not used in English, but which provide the basis of the atomic symbol, or have entered into chemical nomenclature, or are IUPAC-approved alternatives, are added in parentheses in Table I.

I-3.3.3   Atoms of atomic number less than 101 and their symbols

For use in chemical formulae each atom is represented by a unique symbol in upright type as shown in Table I. In addition, the symbols D and T may be used for the hydrogen isotopes of mass numbers two and three, respectively (see Section I-3.5.2). The symbols Id and Va may be used for iodine and vanadium, respectively, only when the single letter symbols are inconvenient or inappropriate, e.g., when they might be confused with Roman numerals.

I-3.3.4   **Atoms of atomic number greater than 100**

Elements of atomic numbers greater than 103 are often referred to in the scientific literature but receive names only after they have been 'discovered'. Names are needed for these elements even before their existence has been established and therefore IUPAC has approved a systematic nomenclature and series of three-letter symbols for the atoms of such elements (see Section I-3.3.5 and Table II).

The existence of this systematic nomenclature does not override the right of discoverers of new elements to suggest other names to IUPAC after their claim has been established beyond all doubt in the general scientific community. For elements 101–103 the systematic names are minor alternatives to the other names recommended by IUPAC. The status of these trivial names and symbols is in no way affected by the recommendation of systematic names for elements of atomic number greater than 100. The systematic names and symbols for elements of atomic number greater than 103 are the only approved names and symbols for those elements until other names are accepted by the Commission on the Nomenclature of Inorganic Chemistry (see Section I-3.3.2).

I-3.3.5   **Systematic nomenclature and symbols of atoms of atomic number greater than 100**

The name is derived directly from the atomic number of the element using the following numerical roots.

0 = nil	3 = tri	6 = hex	9 = enn
1 = un (see Note 3a)	4 = quad	7 = sept	
2 = bi	5 = pent	8 = oct	

The roots are put together in the order of the digits which make up the atomic number and terminated by 'ium' to spell out the name (see Note 3b). The final 'n' of 'enn' is elided when it occurs before 'nil', and the final 'i' of 'bi' and of 'tri' when it occurs before 'ium'. The symbol of an element of atomic number greater than 103 is composed of the initial letters of the numerical roots which make up the name.

Some names thus derived and the other IUPAC-approved names of elements 101, 102, and 103 are listed in order of atomic number in Table II together with their IUPAC-approved symbols. Those of atomic numbers 101 to 109 are also listed in Table I.

I-3.4   **INDICATION OF MASS, CHARGE, AND ATOMIC NUMBER USING INDEXES (SUBSCRIPTS AND SUPERSCRIPTS)**

Mass, ionic charge, and atomic number of a nuclide are indicated by means of three indexes (subscripts and superscripts) placed around the symbol. The positions are occupied thus:

    left upper index   mass number
    left lower index   atomic number
    right upper index   ionic charge

Ionic charge on atoms of symbol A is indicated by A^{n+} or A^{n-}, not by A^{+n} or A^{-n}.

Note 3a. The root 'un' is pronounced with a long 'u', to rhyme with 'moon'.
Note 3b. In the element names each root is to be pronounced separately.

Examples:

$^{32}_{16}S^{2+}$ represents a doubly ionized sulfur atom of atomic number 16 and mass number 32.

The nuclear reaction between $^{26}_{12}Mg$ and $^{4}_{2}He$ nuclei to yield $^{29}_{13}Al$ and ^{1}H nuclei is written as below.

$$^{26}Mg(\alpha, p)^{29}Al \qquad \text{(see Note 3c)}$$

For the nomenclature of isotopically modified compounds and the use of atomic symbols to indicate isotopic modification in chemical formulae, see Chapter I-4 (Note 3d).

[The right lower position of an atomic symbol is reserved for an index (subscript) indicating the number of such atoms in a formula, for example, S_8 is the formula of a molecule containing eight sulfur atoms (see Section I-3.6). For precise formalisms when charges or oxidation states are also shown, see Sections I-4.4.1 and I-4.4.2.].

I-3.5 ISOTOPES

I-3.5.1 Isotopes of an element

These all bear the same name (but see Section I-3.5.2). They are designated by mass numbers (see Sections I-3.2.4 and I-3.4). For example, the atom of atomic number 8 and mass number 18 is named 'oxygen-18' and has the symbol ^{18}O.

I-3.5.2 Isotopes of hydrogen

Hydrogen is an exception to the rule in Section I-3.5.1. The three isotopes, ^{1}H, ^{2}H, and ^{3}H, have the names 'protium', 'deuterium', and 'tritium', respectively. For the latter two the symbols D and T may be used but ^{2}H and ^{3}H are preferred, because D and T disturb the alphabetical ordering in formulae (see Chapter I-4).

It is to be noted that these names give rise to the names proton, deuteron, and triton for the cations $^{1}H^{+}$, $^{2}H^{+}$, and $^{3}H^{+}$, respectively. Because the name 'proton' is often used in contradictory senses, of isotopically pure $^{1}H^{+}$ ions on the one hand, and of the naturally occurring undifferentiated isotope mixture on the other, the Commission recommends that the latter mixture be designated generally by the name hydron, derived from hydrogen.

I-3.6 ELEMENTS

I-3.6.1 Name of an element or elementary substance of infinite or indefinite molecular formula or structure

An element of this kind, which could arise, for example, because it is a mixture of allotropes (Section I-3.7), bears the same name as the atom.

Note 3c. See *Quantities, Units and Symbols in Physical Chemistry*, Blackwell Scientific Publications, Oxford, 1987, Section 2.10, General Chemistry, other Symbols and Conventions in Chemistry.

Note 3d. A more extended treatment is presented in Isotopically Modified Compounds, *Pure Appl. Chem.*, **53**, 1887 (1981).

Examples:
1. Cu(solid), copper
2. Na(solid), sodium
3. $S_6 + S_8 + S_n$, sulfur
4. Se_n, selenium

I-3.6.2   **Name of an element or elementary substance of definite molecular formula**

These are named by adding the appropriate numerical prefix (Table III) (mono-, di-, tri-, tetra-, penta-, hexa-, hepta-, octa-, nona-, deca-, undeca-, and dodeca-) to the name of the atom to designate the number of atoms in the molecule. The prefix mono is not used except when the element does not normally exist in a monoatomic state.

Examples:
1. H, monohydrogen   5. O_3, trioxygen
2. N, mononitrogen   6. P_4, tetraphosphorus
3. N_2, dinitrogen   7. S_8, octasulfur
4. Ar, argon

I-3.7   ALLOTROPIC MODIFICATIONS

I-3.7.1   **Allotropes**

Allotropic modifications of an element bear the name of the atom from which they are derived, together with a descriptor to specify the modification. Common descriptors are Greek letters, α, β, γ, etc., colours, and, where appropriate, mineral names, such as graphite and diamond for the well known forms of carbon. Such names should be regarded as provisional common names to be used until structures have been established, after which a rational system based on molecular formula (see Section I-3.7.2) or crystal structure (see Section I-3.7.3) is recommended. The well-established customary names or descriptors continue for the present as allowed alternatives for common structurally defined allotropes of carbon, phosphorus, sulfur, tin, and iron (Examples in Chapter I-6, and in Section I-3.7.3), except when they can be treated under Section I-3.7.2. Customary names will also continue to be used for amorphous modifications of an element and for those which in their commonly occurring forms are mixtures of closely related structures (such as graphite) or have an ill-defined disordered structure (such as red phosphorus) (see Section I-3.7.4).

I-3.7.2   **Allotropic modifications which are constituted of discrete molecules**

Systematic names are based on the number of atoms in the molecule which is indicated by a numerical prefix (see Section I-3.6.2). If the number is great and unknown, as in long chains or large rings, the prefix poly- may be used. Where necessary, appropriate prefixes of Table V may be used to indicate structure. When it is desired to specify a particular polymorphous form of a molecular element of definitive structure (such as S_8 in α- or β-sulfur) the method of Section I-3.7.3 should be used.

Examples:

Symbol	Trivial name	Systematic name
1. H	atomic hydrogen	monohydrogen
2. O_2	oxygen	dioxygen
3. O_3	ozone	trioxygen
4. P_4	white phosphorus (yellow phosphorus)	tetraphosphorus
5. S_6	—	hexasulfur
6. S_8	α-sulfur, β-sulfur	octasulfur
7. S_n	μ-sulfur (plastic sulfur)	polysulfur

I-3.7.3   **Crystalline allotropic modifications of an element**

These are polymorphs of the elements. Each can be named by adding in parentheses after the name of the atom [element] the Pearson symbol (see Note 3e) which defines the structure of the allotrope in terms of its Bravais lattice (crystal class and type of unit cell) and number of atoms in its unit cell (Table I-3.1 and Chapter I-6). Thus, 'iron ($cF4$)' is that allotropic modification of iron (γ-iron) with a cubic (c), all-face-centred (F) lattice

Table I-3.1 Pearson symbols used for the fourteen Bravais lattices

System	Lattice symbol[a]	Pearson symbol
Triclinic	P	aP[b]
Monoclinic	P	mP
	S[c]	mS
Orthorhombic	P	oP
	S	oS
	F	oF
	I	oI
Tetragonal	P	tP
	I	tI
Hexagonal (and trigonal P)	P	hP
Rhombohedral	R	hR
Cubic	P	cP
	F	cF
	I	cI

[a] P, S, F, I, and R are primitive, one-face-centred, all-face centred, body-centred, and rhombohedral lattices, respectively. The letter C was formerly used in place of S (see also Section I-6.5).
[b] The letter 'a' is used to designate the triclinic lattice because 't' is pre-empted by tetragonal; 'a' is taken from a superseded synonym for triclinic, namely anorthic.
[c] Second setting, y-axis unique.

Note 3e. W. B. Pearson, *Lattice Spacings and Structures of Metals*, Vol. 2, Pergamon Press, Oxford, 1967, pp. 1, 2. For tabulated lattice parameters and data on elemental metals and semi-metals, see pp. 79–91.

containing four atoms of iron in the unit cell. The extensive trivial nomenclature of allotropes is discouraged.

In a few cases, the Pearson symbol fails to differentiate between two crystalline allotropes of the element. In such event the space group is added to the parentheses (Note 3f). Alternatively, a notation involving compound type may be useful (see Chapter I-6).

Examples:

Symbol	Trivial name	Systematic name
1. P_n	black phosphorus	phosphorus ($oC8$)
2. C_n	diamond	carbon ($cF8$)
3. C_n	graphite (common form)	carbon ($hP4$)
4. C_n	graphite (less common form)	carbon ($hR6$)
5. Fe_n	α-iron	iron ($cI2$)
6. Fe_n	γ-iron	iron ($cF4$)
7. Sn_n	α- or grey tin	tin ($cF8$)
8. Sn_n	β- or white tin	tin ($tI4$)
9. Mn_n	α-manganese	manganese ($cI58$)
10. Mn_n	β-manganese	manganese ($cP20$)
11. Mn_n	γ-manganese	manganese ($cF4$)
12. Mn_n	δ-manganese	manganese ($cI2$)
13. S_6	—	sulfur ($hR18$)
14. S_{20}	—	sulfur ($oP80$)

I-3.7.4   Solid amorphous modifications and commonly recognized allotropes of indefinite structure

These are distinguished by customary descriptors such as a Greek letter, or names based on physical properties, or mineral names (see Examples in Section I-3.7.3).

Examples:

1. C_n vitreous carbon
2. C_n graphitic carbon [carbon in the form of graphite, irrespective of structural defects]
3. P_n red phosphorus [a disordered structure containing parts of phosphorus ($oC8$) and parts of tetraphosphorus]
4. As_n amorphous arsenic

I-3.8   GROUPS OF ELEMENTS

I-3.8.1   Groups of elements in the Periodic Table and their subdivisions

The numbering of the Groups of atoms in the Periodic Table from Group I to Group VIII is well established internationally, but the subdivision of these Groups into Typical

Note 3f. For example, the two forms of Se, α-selenium and β-selenium, both Se_n, are distinguished by the symbols ($mP32$, $P2_1/n$) and ($mP32$, $P2_1/a$), respectively.

Table I-3.2 Designation of Groups in the Periodic Tables*

Groups	1*	2	3	4	5	6	7	8	9	10	11	12	13	14	15	16	17	18
	[H]†																	He
	Li	Be											B	C	N	O	F	Ne
	Na	Mg											Al	Si	P	S	Cl	Ar
	K	Ca	Sc	Ti	V	Cr	Mn	Fe	Co	Ni	Cu	Zn	Ga	Ge	As	Se	Br	Kr
	Rb	Sr	Y	Zr	Nb	Mo	Tc	Ru	Rh	Pd	Ag	Cd	In	Sn	Sb	Te	I	Xe
	Cs	Ba	La-Lu‡	Hf	Ta	W	Re	Os	Ir	Pt	Au	Hg	Tl	Pb	Bi	Po	At	Rn
	Fr	Ra	Ac-Lr§	Unq**	Unp	Unh	Uns	Uno	Une	Uun								

*System of numbering proposed to overcome the disparate international use of A and B to designate the sub-Groups (See Section I-3.8.1 and Appendix).

† H is anomalous, and may also be considered as an element of Group 17.

‡ The lanthanoids (see Section I-3.8.2)

§ The actinoids (see Section I-3.8.2)

** It is assumed that these elements of atomic number greater than 103 will fall in the Groups indicated.

Elements and A and B sub-Groups has received disparate interpretation and use in different parts of the world. Consequently, this usage is to be avoided. The recommendations made in Table I-3.2 and the frontispiece to this volume are those which the Commission, after extensive consultation, judges to be most clear and straightforward (Note 3g). They differ from those made in the second edition (1970) of the *Nomenclature of Inorganic Chemistry*. The elements (except hydrogen) of Groups 1, 2, 13, 14, 15, 16, 17, and 18 are those designated as Main Group elements, and except in Group 18 the first two elements of each Main Group are termed Typical Elements. The elements of Groups 3–11 are Transition Elements. Optionally the letters, s, p, d and f may be used to distinguish different blocks of elements. If appropriate for a particular purpose, the various Groups may be named from the first element in each as underlined in Table I-3.2, for example, elements of the boron Group [B, Al, Ga, In, Tl], elements of the titanium Group [Ti, Zr, Hf, Unq], etc.

I-3.8.2 **Collective names of groups of like atoms**.

The following collective names for groups of atoms are IUPAC-approved: actinoids or actinides (Ac, Th, Pa, U, Np, Pu, Am, Cm, Bk, Cf, Es, Fm, Md, No, Lr), lanthanoids or lanthanides (La, Ce, Pr, Nd, Pm, Sm, Eu, Gd, Tb, Dy, Ho, Er, Tm, Yb, Lu) (Note 3h), alkaline earth metals (Ca, Sr, Ba, Ra), chalcogens (O, S, Se, Te, Po), halogens (F, Cl, Br, I, At) (Note 3i), noble gases (He, Ne, Ar, Kr, Xe, Rn), alkali metals (Li, Na, K, Rb, Cs, Fr), and rare earth metals (Sc, Y and the lanthanoids).

The collective name 'pnicogens' has been proposed for the group of atoms of N, P, As, Sb and Bi, but is not IUPAC-approved.

A transition element is an element whose atom has an incomplete d-sub-shell, or which gives rise to a cation or cations with an incomplete d-sub-shell. The First Transition Series of elements is Sc, Ti, V, Cr, Mn, Fe, Co, Ni, Cu. The Second and Third Transition Series are similarly derived: these include the lanthanoids and actinoids, respectively. The latter are designated inner (or f) transition elements of their respective Periods in the Periodic Table.

Because of the inconsistent usage in different languages, the word 'metalloid' should be abandoned. Elements should be classified as metals, semi-metals and non-metals.

Note 3g. For a discussion of other forms and notations for the Periodic Table, consult the Appendix to this volume.

Note 3h. Although actinoid means 'like actinium' and so should not include actinium, actinium has become included by common usage. Similarly, lanthanoid. The ending -ide normally indicates a negative ion, and therefore 'lanthanoid' and 'actinoid' are preferred to 'lanthanide' and 'actinide'. However, owing to wide current use, 'lanthanide' and 'actinide' are still allowed.

Note 3i. The generic terms chalcogenide and halogenide (or halide) are used in naming compounds of the elements in question.

I-4   Formulae

CONTENTS

I-4.1   INTRODUCTION

Formulae (empirical, molecular, and structural) provide a simple and clear method of designating compounds. They are of particular importance in chemical equations and in descriptions of chemical procedures. In order to avoid ambiguity and for many other purposes, e.g., use in machine documentation systems, standardization is useful (Note 4a).

I-4.2   DEFINITIONS OF TYPES OF FORMULA

I-4.2.1   **Empirical formula**

The empirical formula of a compound is formed by juxtaposition of the atomic symbols with their appropriate subscripts (Section I-3.4) to give the simplest possible formula expressing the composition. For the order of citation of symbols see Section I-4.6, but, *in the absence of any other ordering criterion*, the alphabetical order of symbols should be used in an empirical formula except that for compounds containing organic groups, C and H are usually cited first and second, respectively. Molecular or ionic masses cannot be calculated from an empirical formula.

Examples:

1. ClK
2. CaO_4S
3. O_4Rb_2S
4. Cl_6K_3Mo
5. $ClHg$
6. NS
7. $C_6FeK_4N_6$
8. $C_{10}H_{10}ClFe$
9. $Cl_6H_8N_2W$
10. $BrClH_3N_2NaO_2Pt$

I-4.2.2   **Molecular formula**

The molecular formula, for compounds consisting of discrete molecules, is a formula according with the relative molecular mass (or the structure). For the order of citation of symbols, see Section I-4.6.

Examples:

1. S_2Cl_2 (not SCl)
2. $H_4P_2O_6$ (not H_2PO_3)
3. Hg_2Cl_2 (not $HgCl$)

When the relative molecular mass varies with temperature or other conditions, the empirical formula is preferred except where it is desirable to indicate the molecular complexity.

Examples:

4. S 5. P 6. NO_2 (and not S_8, P_4, N_2O_4)

For criteria for ordering of symbols, see Section I-4.6.

Note 4a. The general use of formulae in texts is not recommended, but in cases where bulky and cumbersome names would occur a formula may be preferable, except at the beginning of a sentence.

In referring to ions, radicals, etc., rather than molecular species, some authorities prefer the general term 'group formula'.

I-4.2.3   **Structural formula**

A structural formula (such as a stereoformula, Note 4b) gives information about the way the atoms in a molecule are connected and arranged in space. A line formula may contain structural information, and a displayed formula (as in Example 2, below) probably more. If needed, it can include structural prefixes (see Table V and Section I-4.5.3). The recommendations for presenting molecular formulae may no longer be valid if structural information is inserted.

Examples:

1.
$$\left[\begin{array}{ccc} O & O & O \\ OP{-}O{-}P{-}O{-}PO \\ O & O & O \end{array} \right]^{5-}$$

2.
$$\left[\begin{array}{c} (C_2H_5)_3Sb \diagdown \\ \qquad Pt \diagdown \\ (C_2H_5)_3Sb \diagup \quad \diagup I \\ \qquad \qquad I \end{array} \right]$$

3. $[(NH_3)_5Cr{-}OH{-}Cr(NH_3)_5]Cl_5$

4. $[Pt(\overline{NH_2CH_2CH_2NH_2})Cl_2]$

I-4.2.4   **Solid state structural information**

Structural information can also be given by indicating structural type as a qualification of a molecular formula (see Section I-6.7.2). For example, polymorphs may be indicated by adding in parentheses an abbreviated expression for the crystal system. Structures may also be designated by adding the name of a type-compound in italics in parentheses, but such usage may not be unambiguous. There are at least ten varieties of ZnS(h). Where there are a number of polymorphs crystallizing in the same crystal system, these may be differentiated by the Pearson symbol (see also Section I-3.7.3). Greek letters are frequently employed to designate polymorphs, but the usage is often confused and contradictory. Consequently, one of the methods mentioned here is generally preferable.

Examples:

1. $TiO_2(t)$ (*rutile* type)
2. $TiO_2(t)$ (*anatase* type)
3. $TiO_2(r)$ (*brookite* type)
4. AuCd(c), or AuCd(*CsCl* type)

Note 4b. See the *Nomenclature of Organic Chemistry*, 1979 edition, pp. 484–6.

I-4.3   INDICATION OF PROPORTIONS OF CONSTITUENTS

I-4.3.1   **Numbers of atoms or groups**

The number of identical atoms or atomic groups in a formula is indicated by an arabic numeral as right lower index (subscript) of the symbol or symbols it qualifies (Section I-3.4). The symbol may be in parentheses ( ), square brackets [ ], or braces { }, or stand alone. Enclosing marks may be used to separate symbols, if ambiguities might otherwise arise.

Examples:

1. $CaCl_2$
2. $[Co(NH_3)_6]_2(SO_4)_3$
3. $Ca_3(PO_4)_2$
4. $[\{Fe(CO)_3\}_3(CO)_2]^{2-}$
5. $K[Os(N)O_3]$

Solvating molecules and similar bound molecules in addition compounds (such as may occur in Lewis acid–base adducts and in charge-transfer molecular complexes) are not indicated as coordination compounds unless they can be unambiguously designated as such. Their proportions may be indicated by arabic numerals preceding their formulae. The formulae of the constituent compounds are separated by a centre dot. For particular classes of compound there may be specific ordering rules for formulae (see, e.g., Section I-4.6.8).

Examples:

6. $Na_2CO_3 \cdot 10H_2O$
7. $8H_2S \cdot 46H_2O$
8. $NH_3 \cdot B(CH_3)_3$

I-4.3.2   **Solid state phases**

For the formulae of solid solutions and non-stoichiometric phases, see Chapter I-6.

I-4.4   INDICATION OF OXIDATION STATE AND CHARGE OF CONSTITUENTS

I-4.4.1   **Oxidation states**

The oxidation state of an element in a formula may be indicated by a number written as right upper index (superscript) in roman numerals. In addition, oxidation state zero may be represented by the numeral 0. If an element occurs with more than one oxidation state in the same formula, the element symbol is repeated, each symbol being assigned a number cited in sequence in increasing magnitude and from negative to positive.

Examples:

1. $[P^V_2Mo_{18}O_{62}]^{6-}$
2. $K[Os^{VIII}(N)O_3]$
3. $[Mo^V_2Mo^{VI}_4O_{18}]^{2-}$
4. $Pb^{II}_2Pb^{IV}O_4$
5. $[Os^0(CO)_5]$
6. $Na_2O^{-I}_2$

Where it is not feasible or reasonable to define an oxidation state for each individual member of a group (or cluster), it is recommended that the overall oxidation level of the group be defined by a formal ionic charge, indicated as in Section I-4.4.2. This avoids the use of fractional oxidation states (Note 4c).

Examples:

7. O_2^- 8. $Fe_4S_4^{3+}$

I-4.4.2   **Ionic charge**

The ionic charge is indicated by means of a right upper index as A^{n+} or A^{n-} (not A^{+n} or A^{-n}) (see Section I-3.4). In cases such as coordination ions and extended structures, $n+$ or $n-$ is used as a right upper index after the appropriate brackets or parentheses. Note the general presentation when brackets are not used, $X_x Y_y^{n+}$, etc., and not $X_x Y_y^{n+}$.

Examples:

1. Cu^+ 9. HF_2^-
2. Cu^{2+} 10. CN^-
3. NO^+ 11. $S_2O_7^{2-}$
4. $[Al(H_2O)_6]^{3+}$ 12. $[Fe(CN)_6]^{4-}$
5. $H_2NO_3^+$ 13. $(O_3POSO_3)^{3-}$
6. $[PCl_4]^+$ 14. $[PW_{12}O_{40}]^{3-}$
7. H^- 15. $[(CuCl_3)_n]^{m-}$
8. As^{3-}

I-4.4.3   **Free radicals**

IUPAC recommends that the use of the word radical be restricted to species conventionally termed free radicals. In the current context, a radical is considered to be an atom or group of atoms containing one or more unpaired electrons (see Section I-8.4.1). A radical may also be charged, positively or negatively. In transition metal compounds, it is not often considered necessary to show the presence of unpaired electrons and they are usually not indicated. Even in non-transition-metal radicals the unpaired electrons are rarely shown formally. Nevertheless, any charge on the radical must be explicit.

When it is desired, a radical is indicated by a dot as right superscript to the symbol of the element or group (Note 4d). The formulae of polyatomic radicals are placed in parentheses and the dot is placed as a right superscript to the parentheses. In radicals which are treated as coordination compounds (see Chapter I-10) the dot is a right superscript on the square brackets. The superscript dot(s) merely indicate the presence of

Note 4c. Oxidation state is a formal device for assigning electrons to particular atoms or groups of atoms in a molecule, and for understanding redox reactions. The original rules for determining oxidation states do not allow fractional values. Because it is rarely possible to assign electrons to specific atoms of groups where fractional oxidation states might be postulated, it seems better to reject the device of fractional oxidation state in favour of the charge formalism suggested here.

Note 4d. This use differs in detail from the practice recommended in the *Nomenclature of Organic Chemistry*, 1979 edition, Rule C-81, and in solid-state chemistry (Chapter I-6). The use of the dot in structural formulae in which the position of the unpaired electron needs to be specified should follow the practice of Rule C-81 of the *Nomenclature of Organic Chemistry*, 1979 edition.

unpaired electron(s), and not their location. In radical ions the dot precedes the charge (Note 4e).

Examples:

1. $H^{\bullet}$	8. $(CN)^{\bullet}$	15. $(O_2)^{\bullet -}$
2. $Br^{\bullet}$	9. $[Mn(CO)_5]^{\bullet}$	16. $(NO)^{\bullet -}$
3. $^7Li^{\bullet}$	10. $(HgCN)^{\bullet}$	17. $(Cl_2)^{\bullet -}$
4. $^{23}Na^{\bullet}$	11. $(SnCl_3)^{\bullet}$	18. $(SO_2)^{\bullet -}$
5. $(NO_2)^{\bullet}$	12. $[V(CO)_6]^{\bullet}$	19. $(B\underset{\bullet}{H_3})^{\bullet -}$
6. $(OH)^{\bullet}$	13. $(Ag_2)^{\bullet +}$	20. $(CO_2)^{\bullet -}$
7. $(PO)^{\bullet}$	14. $(NH_3)^{\bullet +}$	21. $[Cr(C_{10}H_7)_2]^{\bullet +}$
		22. $[Cr(C_{10}H_7)_2]^{\bullet 3-}$

Two unpaired electrons may be indicated by two dots or $2^{\bullet}$. More than two should always be indicated by superscript $n^{\bullet}$, where $n \geq 3$.

Examples:

23. $(O_2)^{\bullet\bullet}$ or $(O_2)^{2\bullet}$
24. $(N_2)^{\bullet\bullet 2-}$ or $(N_2)^{2\bullet 2-}$
25. $(N_2O)^{\bullet\bullet 2+}$ or $(N_2O)^{2\bullet 2+}$
26. $[FeCl_4]^{4\bullet 2-}$

In structural formulae, it may be of value to use the dot to indicate the location of the unpaired electron(s).

I-4.5   FURTHER MODIFIERS OF FORMULAE

I-4.5.1   Optically active compounds

The sign of optical rotation is placed in parentheses, the wavelength (in nm) being indicated as a right subscript. The whole symbol is placed before the formula and refers to the sodium D-line unless otherwise stated (Note 4f).

Examples:

1. $(+)_{589}[Co(NH_2CH_2CH_2NH_2)_3]Cl_3$
2. $(-)_{589}[Co\{(-)NH_2CH(CH_3)CH_2NH_2\}_3]Cl_3$

I-4.5.2   Excited states

Excited electronic states may be indicated by an asterisk as right superscript. This practice does not differentiate between different excited states.

Note 4e. This recommendation is consistent with the *Nomenclature of Organic Chemistry*, 1979 edition, Rules C-83.3 and C-84.4 and the *Compendium of Chemical Terminology*, but is at variance with the recommendations for mass spectrometry, *Pure Appl. Chem.*, **50**, 56 (1978). Where it is considered that a representation such as $M^{2\cdot 2-}$ might be confusing, the use of parentheses is suggested, namely $M^{(2\cdot)(2-)}$.

Note 4f. The practice is described more fully in the *Nomenclature of Organic Chemistry*, 1979 edition, p. 480, Rule E-4.4.

Examples:
1. He*
2. NO*

I-4.5.3 **Structural modifiers**

Modifiers such as *cis-*, *trans-*, etc., are listed in Table V. Usually such modifiers are used as italicized prefixes and are connected to the formula by a hyphen.

Examples:
1. *cis*-[PtCl$_2$(NH$_3$)$_2$]
2. *trans*-[PtCl$_4$(NH$_3$)$_2$]
3. *trans*-[MoCl$_4$(thf)$_2$] · *trans*-[ReCl(N$_2$)(PMe$_2$Ph)$_4$]

Example 3 contains the abbreviation thf for tetrahydrofuran. Table I-10.5 lists common abbreviations, but see also Section I-4.6.7.2.

The modifier μ for designating an atom or group linking coordination centres is often encountered and it may be used as an infix. The bridging group(s) are normally placed last in the formula, modified by the prefix μ and enclosed by parentheses, and in any case preferably after the central atom(s) in the formula. If the bridging group spans more than two atoms, μ may itself be modified, μ_3, μ_4, ..., μ_n. If there is more than one bridging group, the bridges should be cited in increasing order of bridge multiplicity. If two or more bridging groups have the same multiplicity, then the alphabetical order of the various bridge symbols decides the order of citation. See Chapter I-2 and particularly Chapter I-10 for a full discussion of coordination compounds and the uses of structural modifiers.

Examples:
4. [{Cr(NH$_3$)$_5$}$_2$(μ-OH)]Cl$_5$
5. [Cr$_3$(μ-CH$_3$COO)$_6$(μ_3-O)]Cl
6. [{Co(CN)$_5$}{Fe(CN)$_5$}(μ-CN)]

I-4.6 SEQUENCE OF CITATION OF SYMBOLS

I-4.6.1 **Priorities**

I-4.6.1.1 *General*

The sequence of symbols in a formula is always arbitrary and in any particular case should be a matter of convenience. For example, the sequence used in an index will be appropriate to its own requirements. Where there are no over-riding requirements, then the guidelines below should be used; they are summarized on the following page in Table I-4.1.

I-4.6.1.2 *Electronegativities and citation order*

In a formula the order of citation of symbols is based upon *relative* electronegativities, the more electropositive constituent(s) being cited first. In the absence of any universal

Table I-4.1 Assignment of formulae of compounds

(i) Assign symbols to the constituents (Chapter I-3).

(ii) Indicate proportions of constituents (I-4.3).

(iii) Divide constituents into electropositive and electronegative (I-4.6.1.2). This requires decisions concerning compound type. There are special rules for acids (I-4.6.1.2), polyatomic groups (I-4.6.5), binary compounds (I-4.6.2), chain compounds (I-4.6.3), intermetallic compounds (I-4.6.6), coordination compounds (I-4.6.7), and addition compounds (I-4.6.8).

(iv) Assemble the formula (I-4.6).

(v) Insert appropriate modifiers (geometrical, etc.) (I-4.2.3, I-4.2.4, I-4.5, I-4.7).

(vi) Insert oxidation states, charges, etc., if required (I-4.4).

---

electronegativity scale for this purpose, Table IV should be used as a guide to relative electronegativities. In general, elements before Al in the Table IV sequence should be regarded as electronegative, those after B as electropositive. This is not to be regarded as prescriptive. In binary compounds between non-metals (Section I-4.6.2), in polyatomic groups (Section I-4.6.5), and in intermetallic compounds (Section I-4.6.6), the constituent cited first is treated as if it were electropositive, even when general criteria would suggest that it is not. In any case, specific examples may require modification of these recommendations, as in the halogen oxides. In the formulae of Brønsted acids (see Section I-5.3.2), acid hydrogen is considered to be an electropositive constituent and immediately precedes the anionic constituents.

If the compound contains more than one electropositive or electronegative constituent, the sequence within each class is the alphabetical order of their symbols.

Examples:

1. KCl
2. $CaSO_4$
3. HBr
4. H_2SO_4
5. $[Cr(H_2O)_6]Cl_3$
6. $H[AuCl_4]$
7. $Na_2B_4O_7$
8. $NaHSO_4$
9. $IBrCl_2$
10. O_2ClF_3

I-4.6.1.3 *Alphabetical order*

A single letter symbol, e.g., B, always precedes a two-letter symbol with the same initial letter, e.g., Be. Non-replaceable hydrogen in an anion is a ligand taking its normal alphabetical position. Replaceable (acid) hydrogen is treated as described in Section I-4.6.1.2. NH_4 is treated as a single symbol and is listed after Ne. If the first symbols are the same, a monoatomic constituent precedes a polyatomic or complex one, e.g., O^- precedes OH^-. Where the entities to be arranged in a formula are polyatomic, then the order of citation is decided by selecting a particular atomic symbol to characterize the entity. The first symbol in the formula of a coordination entity or polyatomic group, as written according to Sections I-4.6.2–I-4.6.7, determines the alphabetical order. For example, SCN, UO_2, NO_3, OH, and $[Zn(H_2O)_6]^{2+}$ are ordered under S, U, N, O, and Zn, respectively. See also Chapter I-2.

If the first symbols are the same, the symbol with the lesser right index is cited first, e.g., NO_2 precedes N_2O_2. If this still does not discriminate, the subsequent symbols are used alphabetically and numerically to define the order, e.g., NO_2 precedes NO_3, NH_2 precedes NO_2. The order of citation of some anions of nitrogen is the following: N^{3-}, NH_2^-, NO_2^-, NO_3^-, $N_2O_2^{2-}$, N_3^-.

The recommendations contained in Sections I-4.6.1.2 and I-4.6.1.3 are illustrated in the Examples below.

Examples:

1. $KMgF_3$
2. $NaTl(NO_3)_2$
3. $CaTiO_3$
4. $TiZnO_3$
5. $NaNH_4HPO_4 \cdot 4H_2O$
6. LiH_2PO_4
7. KHS
8. $NaHPHO_3$
9. $K_2Mg_2V_{10}O_{28} \cdot 16H_2O$
10. $Na(UO_2)_3[Zn(H_2O)_6](C_2H_3O_2)_9$
11. $MgCl(OH)$
12. $VOSO_4$
13. $AlLiMn^{IV}_2O_4(OH)_4$
14. $FeO(OH)$
15. ZnN_3OH
16. $Co_2NO_3(OH)_3$
17. $Na_6ClF(SO_4)_2$
18. $Pb(OH)(C_2H_3O_2)$

Deviation from alphabetical order is allowed to emphasize similarities between compounds.

Examples:

19. $CaTiO_3$ and $ZnTiO_3$

I-4.6.2 Binary compounds between non-metals

In accordance with established practice, in binary compounds between non-metals that constituent is placed first which appears earlier in the following sequence (Note 4g).

Rn, Xe, Kr, Ar, Ne, He, B, Si, C, Sb, As, P, N, H, Te, Se, S, At, I, Br, Cl, O, F.

Examples:

1. NH_3
2. H_2S
3. Cl_2O
4. OF_2
5. B_2H_6
6. PH_3

I-4.6.3 Chain compounds

For chain compounds containing three or more different elements, the sequence should generally be in accordance with the order in which the atoms are actually bound in the molecule or ion.

Examples:

1. $-SCN$ (not CNS)
2. $HOCN$ (cyanic acid)
3. $HONC$ (fulminic acid)
4. $(O_3POSO_3)^{3-}$

Note 4g. Antimony is here classified as a non-metal, but its properties are intermediate between metals and non-metals. In other contexts it may be considered to be a metal. The sequence quoted here is a modified electronegativity sequence.

I-4.6.4 Polyatomic ions

Polyatomic ions, whether complex or not, are treated in a similar fashion. The central atom(s) (e.g., I in $[ICl_4]^-$, U in UO_2^{2+}, Si and W in $[SiW_{12}O_{40}]^{4-}$) or characteristic atom (e.g., Cl in ClO^-, O in OH^-) is cited first and then the subsidiary groups follow in alphabetical order of the symbols in each class.

Examples:

1. SO_4^{2-}
2. UO_2^{2+}
3. NO_2^-
4. ClO^-
5. OH^- (Note 4h)
6. $P_2O_7^{4-}$
7. $[P_2W_{18}O_{62}]^{6-}$
8. $[Mo_6O_{18}]^{2-}$
9. $[CrO_7S]^{2-}$
10. $[SiW_{12}O_{40}]^{4-}$
11. $[BH_4]^-$
12. $[ICl_4]^-$

I-4.6.5 Polyatomic compounds or groups

It is necessary to define the central atom of the compound or group, and this is always cited first. If two or more different atoms or groups are attached to a single central atom, the symbol of the central atom is followed by the symbols of the remaining atoms or groups in alphabetical order. The sole exceptions are acids, in the formulae of which hydrogen is placed first. When part of the molecule is a group, such as $>P=O$, which occurs repeatedly in a number of different compounds, these groups may be treated as forming the positive part of the compound.

Examples:

1. $PBrCl_2$
2. $SbCl_2F$
3. $POCl_3$ or PCl_3O
4. $PSCl_3$ or PCl_3S
5. H_3PO_4

I-4.6.6 Intermetallic compounds

The constituents (including Sb) are placed in the alphabetical order of their symbols. Deviation from this order is allowed, for example to emphasize ionic character or when compounds with analogous structures are compared, but unless there are over-riding considerations, the alphabetical order is to be followed.

Examples:

1. Au_2Bi
2. $NiSn$
3. Mg_2Pb
4. Cu_5Zn_8 } (analogous structures)
5. Cu_5Cd_8 }
6. Na_3Bi_5 (ionic character)

Note 4h. The hydroxide ion is represented by the symbol OH^-, although the recommendations for the formulae of acids (see Section I-4.6.1.2 and Chapter I-9) would suggest HO^-. Example 5 accords with majority practice.

I-4.6.7   Coordination compounds

I-4.6.7.1   *Summary of formalisms used*

Chapter I-10 deals with the nomenclature of coordination compounds. In formulae, they are treated as are other polyatomic groups. In the formula of a coordination entity, the symbol of the central atom(s) is placed first, followed by the ionic and then the neutral ligands. The order of citation of central atoms is decided by use of the alphabetical order of the element symbols. Ligands are cited alphabetically within each class according to the first symbol of their formulae as written according to Sections I-4.6.1–I-4.6.5. Thus, H_2O, NH_3, SiH_3^-, NO_3^-, SO_4^{2-}, and OH^- are cited at H, N, Si, N, S, and O, and ligands containing carbon and hydrogen only are cited under C. Organic ligands containing heteroatoms (atoms other than carbon and hydrogen) have their formulae cited according to the organic system in which C and then H precede other atoms which follow them in alphabetical order. The position of such a formula in the citation sequence is determined by the alphabetical sequence of heteroatom(s). Of two ligands with the same defining atom, that with fewer such atoms precedes that with more. If the numbers of defining atoms are equal, *subsequent* symbols define the sequence. For example, $P(C_2H_5)_3$ and C_5H_5N are cited at P and N; C_5H_5N precedes NH_3; and $C_2H_8N_2$ precedes $C_{10}H_8N_2$.

Square brackets are used to enclose the whole coordination entity whether charged or not. This practice need not be used for simple species such as the common oxoanions, (NO_3^-, NO_2^-, SO_4^{2-}, OH^-, etc.). Enclosing marks are nested within the square brackets as follows: [( )], [{( )}], [{[( )]}], [{{[( )]}}], etc.

A structural formula of a ligand occupies the same place in a sequence as would its molecular formula.

The molecules NO and CO linked to a metal atom are regarded as neutral ligands for nomenclature purposes.

Examples:

1. $K_3[Fe(CN)_6]$
2. $K_2[OsCl_5N]$
3. $[Co(N_3)(NH_3)_5]SO_4$
4. $[Al(OH)(H_2O)_5]^{2+}$
5. $K_2[Cr(CN)_2O_2(O_2)NH_3]$
6. $[Ru(NH_3)_5(N_2)]Cl_2$
7. *cis*-$[PtCl_2\{P(C_2H_5)_3\}_2]$
8. $Na[PtBrCl(NO_2)NH_3]$
9. $[PtCl_2(C_5H_5N)NH_3]$
10. $[Co(C_2H_8N_2)_2(C_{10}H_8N_2)]^{3+}$

I-4.6.7.2   *Abbreviations*

These may be used to represent ligands in formulae, and they are cited in the same place as the formulae they stand for. The abbreviations should be lower case, and enclosed in parentheses. Some commonly used abbreviations are in Tables I-10.5 and Table X.

Examples:

1. $[Pt(py)_4][PtCl_4]$
2. $[Fe(en)_3][Fe(CO)_4]$
3. $[Co(en)_2(bpy)]^{3+}$ or $[Co(C_2H_8N_2)_2(C_{10}H_8N_2)]^{3+}$

The commonly used abbreviations for organic groups, such as Me, Ph, Bu, etc., are acceptable in inorganic formulae. Note that the difference between an anion and its parent acid must be observed. Thus acac is an acceptable abbreviation for acetylacetonate. Acetylacetone (pentane-2,4-dione) then becomes Hacac (see Table I-10.5 and Table X). Failure to observe this convention leads to inconsistencies and the requirement for negative subscripts to show the absence of atoms. Such practices are not recommended.

I-4.6.8   **Addition compounds**

In the formulae of addition compounds the component molecules are cited in order of increasing number; if they occur in equal numbers, they are cited in alphabetical order of the first symbols. Addition compounds containing boron compounds or water are exceptional, in that the water or boron compound is cited last. If both are present, the boron compound precedes water. This is a traditional practice for hydrates.

Examples:
1. $3CdSO_4 \cdot 8H_2O$
2. $Na_2CO_3 \cdot 10H_2O$
3. $Al_2(SO_4)_3 \cdot K_2SO_4 \cdot 24H_2O$
4. $AlCl_3 \cdot 4C_2H_5OH$
5. $8H_2S \cdot 46H_2O$
6. $C_6H_6 \cdot NH_3 \cdot Ni(CN)_2$
7. $2CH_3OH \cdot BF_3$
8. $BF_3 \cdot 2H_2O$

I-4.7   ISOTOPICALLY MODIFIED COMPOUNDS (Note 4i)

I-4.7.1   **General formalism**

The mass number of any specific nuclide can be indicated in the usual way with a left superscript preceding the appropriate atomic symbol (Section I-3.4).

When it is necessary to cite different nuclides at the same position in a formula, the nuclide symbols are written in alphabetical order and when their atomic symbols are identical, then the order is that of increasing mass number. Isotopically modified compounds may be classified as *isotopically substituted* compounds and *isotopically labelled* compounds (Note 4i).

I-4.7.2   **Isotopically substituted compounds**

An isotopically substituted compound has a composition such that all the molecules of the compound have only the indicated nuclide(s) at each designated position. The substituted nuclides are indicated by insertion of the mass numbers as left superscripts preceding the appropriate atom symbols in the normal formula.

Note 4i. For additional material, see Isotopically Modified Compounds, *Pure Appl. Chem.*, **53**, 1887 (1981).

Examples:

1. H^3HO 5. $^{32}PCl_3$
2. $H^{36}Cl$ 6. $K[^{32}PF_6]$
3. $^{235}UF_6$ 7. $K_3{}^{42}K[Fe(^{14}CN)_6]$
4. $^{42}KNa^{14}CO_3$

I-4.7.3   Isotopically labelled compounds

I-4.7.3.1   *Types of labelling*

An isotopically labelled compound may be considered formally as a mixture of an isotopically unmodified compound and one or more analogous isotopically substituted compounds. They may be divided into several different groups. Specifically labelled compounds and selectively labelled compounds are treated here.

I-4.7.3.2   *Specifically labelled compounds*

An isotopically labelled compound is called a specifically labelled compound when a unique isotopically substituted compound is added formally to the analogous isotopically unmodified compound. A specifically labelled compound is indicated by enclosing the appropriate nuclide symbol(s) and multiplying subscript (if any) in square brackets.

Examples:

1. $H[^{36}Cl]$ 4. $[^{13}C]O[^{17}O]$
2. $[^{32}P]Cl_3$ 5. $[^{32}P]O[^{18}F_3]$
3. $[^{15}N]H_2[^2H]$ 6. $Ge[^2H_2]F_2$

I-4.7.3.3   *Selectively labelled compounds*

A selectively labelled compound may be considered as a mixture of specifically labelled compounds. It is indicated by prefixing the formula by the nuclide symbol(s) preceded by any necessary locant(s) (but without multiplying subscripts) enclosed in square brackets.

Examples:

1. $[^{36}Cl]SOCl_2$
2. $[^2H]PH_3$
3. $[^{10}B]B_2H_5Cl$

The number of possible labels for a given position may be indicated by subscripts separated by semicolons added to the atomic symbol(s) in the isotopic descriptor.

Example:

4. $[1\text{-}^2H_{1;2}]SiH_3OSiH_2OSiH_3$

56

I-4.8   FINAL REMARKS

The foregoing recommendations are intended to be guidelines, and to expound the basic principles upon which formulae should be constructed. Table I-4.1 provides guidance in the use of this Chapter for constructing formulae. However, it is often not possible to separate the rules relating to formulae from those relating to names. Consequently, users should refer to more specialized Chapters or documents for information concerning specific types of compound, for example, Chapter I-10 for coordination compounds or see Note 4i for isotopically modified compounds. The principles expounded here are intended to be of general application.

I-5   Names Based on Stoichiometry

CONTENTS

I-5.1   INTRODUCTION

This Chapter provides names for compounds for which little or no structural information is available or for which a minimum of structural information is to be conveyed. In such circumstances, one could base the name on a non-structural empirical formula (Section I-4.2.1). However, in practice a minimum of additional chemical information is known or assumed. Many inorganic compounds are regarded as consisting of electropositive (possibly cationic) and electronegative (possibly anionic) parts. Hence those principles used to assemble symbols in formulae into electropositive and electronegative groups are also used for developing names (see Section I-4.6).

I-5.2   CLASSES AND CITATION OF CONSTITUENTS

In the simplest cases of one electropositive and one electronegative constituent, the names are formed by combining the name of the electropositive constituent with that of the

58

electronegative constituent, suitably qualified by any necessary multiplicative prefixes, and with the former cited first. The two parts of the name are separated by a space (Note 5a). The name may be the simple element name for a monoatomic electropositive constituent, but the names of electronegative constituents are normally anion names (Section I-8.3). The multiplicative prefixes may not be necessary if there is no problem with different oxidation states, even where a strictly rigorous derivation of the name might require them.

Examples:

1. $NaCl$ sodium chloride
2. Ca_3P_2 calcium phosphide
3. Fe_3O_4 triiron tetraoxide
4. Fe_2O_3 diiron trioxide
5. SiC silicon carbide

I-5.3   NAMES OF CONSTITUENTS

I-5.3.1   Electropositive constituents

The name of a monoatomic electropositive constituent is simply the unmodified element name (see Table I). A polyatomic constituent assumes the usual cation name (see Section I-8.2.3 and Chapter I-10), but certain well established radical names (particularly for oxygen-containing species such as nitrosyl and phosphoryl, see Section I-8.4.2.2) are still allowed for specific cases.

Examples:

1. NH_4Cl ammonium chloride
2. UO_2Cl_2 uranyl dichloride
3. $POCl_3$ phosphoryl trichloride
4. $NOHSO_4$ nitrosyl hydrogensulfate
5. $NOCl$ nitrosyl chloride
6. $[Co(NH_3)_6]Br_3$ hexaamminecobalt tribromide
7. OF_2 oxygen difluoride
8. O_2F_2 dioxygen difluoride
9. $O_2[PtF_6]$ dioxygen hexafluoroplatinate

I-5.3.2   Ordering of electropositive constituents

If the compound contains more than one kind of electropositive constituent, the names should be spaced and cited in alphabetical order of the initial letters, or of the second letter if the initial letters are the same, and so on. Thus cobalt precedes copper and potassium precedes praseodymium. Hydrogen is always cited last among electropositive constituents and is separated from the following anion name(s) by a space unless it is

Note 5a. Detailed ordering and spacing practice may differ from language to language.

known to be bound to an anion, when it would be expressed as 'hydrogenphosphate', with a different structural implication from 'hydrogen phosphate' (see Section I-8.5.2).

The sequence of citation of names will vary with the language employed and will often differ from the formula sequence (Chapter I-4). Multiplicative prefixes not an inherent part of the cation name are ignored for the purpose of determining sequence (see Section I-5.5).

Examples:

1.	$KMgCl_3$	magnesium potassium chloride
2.	$NaNH_4HPO_4$	ammonium sodium hydrogen phosphate
3.	$Na(UO_2)_3[Zn(H_2O)_6](C_2H_3O_2)_9$	hexaaquazinc sodium triuranyl nonaacetate
4.	$Cs_3Fe(C_2O_4)_3$	tricaesium iron tris(oxalate) (see Note 5b)
5.	$AlK(SO_4)_2 \cdot 12H_2O$	aluminium potassium bis(sulfate)—water (1/12) or aluminium sulfate—potassium sulfate—water (1/1/24) (Note 5c)
6.	$Ca_3H_3ClF(PO_4)(SO_4)_2$	tricalcium trihydrogen chloride fluoride phosphate bis(sulfate)

I-5.3.3 **Monoatomic electronegative constituents**

The name of a monoatomic electronegative constituent is the element name (Table I) with its ending (-en, -ese, -ic, -ine, -ium, -ogen, -on, -orus, -um, -ur, -y, or -ygen) replaced by the anion designator -ide.

Examples:

1. chloride derived from chlorine
2. carbide derived from carbon
3. tungstide derived from tungsten (Note 5d)
4. arsenide derived from arsenic
5. aluminide derived from aluminium
6. manganide derived from manganese
7. silicide derived from silicon
8. hydride derived from hydrogen
9. nitride derived from nitrogen
10. oxide derived from oxygen
11. sulfide derived from sulfur
12. phosphide derived from phosphorus

There are exceptions to this rule. In certain cases, abbreviated names are preferred. Certain element names, such as those of the noble gases, and names which have endings

Note 5b. The name caesium tris(oxalato)ferrate(III) conveys more knowledge of the structure than the simple name quoted here.

Note 5c. The doubling of the formula is required by the presentation of the second name. The use of the long dash in names of addition compounds is detailed in Section I-5.6.

Note 5d. In those languages which use wolfram for W rather than tungsten, an anion name based on wolframide should be used.

other than those cited above, such as zinc, can be converted to the anion name simply by the addition of the anionic ending -ide, though there may also be a contraction (Note 5e).

Examples:
13. germide-germanium (Note 5f)
14. bismuthide-bismuth
15. cobaltide-cobalt
16. nickelide-nickel
17. zincide-zinc
18. neonide-neon
19. argonide-argon
20. kryptonide-krypton
21. radonide-radon

Finally, there are monoatomic anions whose names in English, though derived as described above, are based on the Latin root of the element names. In these the ending -um or -ium is replaced by -ide.

Examples:
22. argentide-argentum-silver
23. auride-aurum-gold
24. cupride-cuprum-copper
25. ferride-ferrum-iron
26. plumbide-plumbum-lead
27. stannide-stannum-tin
28. natride-natrium-sodium
29. kalide-kalium-potassium

If the compound contains more than one electronegative constituent, then the names of these are ordered alphabetically (compare Section I-5.3.2) on the basis of the initial letters of the names (or subsequent letters if this does not distinguish) ignoring the initial letters of any qualifying multiplicative prefixes. Again, there should be spaces between the constituent names.

Examples:
30. $BBrF_2$ boron bromide difluoride
31. PCl_3O phosphorus trichloride oxide
32. $Na_6ClF(SO_4)_2$ hexasodium chloride fluoride bis(sulfate) (see Section I-5.5.1)
33. $Na_2F(HCO_3)$ sodium fluoride hydrogencarbonate

Note 5e. Noble gas anion names are needed for theoretical discussion and have already appeared in the chemical literature. The ending -on is retained, otherwise radon would have the same root as radium. The root ne- from neon is not obvious, and a root krypt- could lead to kryptate, perhaps to be confused with cryptate.

Note 5f. The systematic name germanide designates an anionic derivative of germane, GeH_4.

Homoatomic electronegative constituents

These have the name of the monoatomic parent, but qualified by a multiplicative prefix, if appropriate (see Section I-5.5.1). It may be necessary to use parentheses to emphasize subtle points of structure.

Examples:
1. Na_4Sn_9 tetrasodium (nonastannide)
2. $TlCl_3$ thallium trichloride (Note 5g)
3. $Tl(I_3)$ thallium (triiodide) (Note 5g)
4. Na_2S_2 sodium disulfide

Some homopolyatomic anions have trivial names which are still allowed.

Examples:
5. O_2^- hyperoxide, or superoxide
6. O_2^{2-} peroxide
7. O_3^- ozonide
8. N_3^- azide
9. C_2^{2-} acetylide

Heteropolyatomic electronegative constituents

The names of these anions take the termination -ate, though a few exceptions are allowed (see Examples 8–19, below). If the anion is more properly considered a coordination entity, then the name is derived using the principles of coordination nomenclature (see Chapter I-10). The name of the anion is then constructed from the names of the ligands and the name of the central or characteristic atom, modified by the ending -ate (see Table I-9.2 for a list of such modified names). In such cases, indication of charge may be necessary to avoid ambiguity.

Examples:
1. $[Cr(NCS)_4(NH_3)_2]^-$ diamminetetrathiocyanatochromate$(1-)$
2. $[Fe(CO)_4]^{2-}$ tetracarbonylferrate$(2-)$
3. $[PF_6]^-$ hexafluorophosphate$(1-)$

The ending -ate is also a characteristic ending for the names of anions of oxoacids and their derivatives (see Chapter I-9). The names sulfate, phosphate, nitrate, etc., are general names for oxoanions containing sulfur, phosphorus, and nitrogen surrounded by ligands, including oxygen, irrespective of their nature and number. The names sulfate, phosphate, and nitrate were originally restricted to the anions of specific oxoacids, namely SO_4^{2-}, PO_4^{3-}, NO_3^-, but this is no longer the case (see Chapter I-9 for a complete discussion).

Note 5g. In Examples 2 and 3, the use of oxidation state designators would be appropriate. See Section I-5.5.2.2.

Examples:

4.	$SO_3{}^{2-}$	trioxosulfate, or sulfite
5.	$SO_4{}^{2-}$	tetraoxosulfate, or sulfate
6.	$NO_2{}^{-}$	dioxonitrate, or nitrite
7.	$NO_3{}^{-}$	trioxonitrate, or nitrate

Many names with -ate endings are still allowed, though they are not completely in accord with the derivations outlined above. Some of these are cyanate, dichromate, diphosphate, disulfate, dithionate, fulminate, hypophosphate, metaborate, metaphosphate, metasilicate, orthosilicate, perchlorate, periodate, permanganate, phosphinate, and phosphonate. For a complete discussion see Chapter I-9.

The exceptional cases where the names end in -ide or -ite rather than -ate are exemplified below (see Section I-8.3.3.8 and Chapter I-9).

Examples:

8.	CN^{-}	cyanide
9.	$NHNH_2{}^{-}$	hydrazide
10.	$NHOH^{-}$	hydroxyamide (Note 5h)
11.	$NH_2{}^{-}$	amide
12.	NH^{2-}	imide
13.	OH^{-}	hydroxide
14.	$AsO_3{}^{3-}$	arsenite
15.	$ClO_2{}^{-}$	chlorite
16.	ClO^{-}	hypochlorite
17.	$NO_2{}^{-}$	nitrite
18.	$SO_3{}^{2-}$	sulfite
19.	$S_2O_4{}^{2-}$	dithionite

I-5.4 ORDER OF CITATION WITHIN CLASSES

The order of citation is based upon the alphabetical order of the names within each class of constituents, electropositive and electronegative (see Section I-4.6). This has been discussed above (Sections I-5.3.1 and I-5.3.2 for electropositive constituents, and Sections I-5.3.3 to I-5.3.5 for electronegative constituents). This means that the order in a name is not necessarily the same as the order of symbols in the corresponding formula, and that the order in a name can change from language to language.

I-5.5 INDICATION OF PROPORTIONS OF CONSTITUENTS

I-5.5.1 Use of multiplicative prefixes

The proportions of the constituents, be they monoatomic or polyatomic, may be indicated by numerical prefixes (mono-, di-, tri-, tetra-, penta-, etc.) as detailed in Table III

Note 5h. This name stems from substitutive nomenclature. The trivial name of this anion is hydroxylamide, and its systematic coordination name is hydroxylimide. Care must be taken to avoid confusion (compare Table I-10.3, Table VIII, and Section I-8.3.3.3, Example 7).

and discussed briefly above (Sections I-5.2 and I-5.3). These precede the names they modify, joined directly without space or hyphen. The final vowels of numerical prefixes should not be elided, except for compelling linguistic reasons. Note that monoxide is an exception. Where the compounds contain elements such that it is not necessary to stress the proportions, for instance, where the oxidation states are usually invariant, then indication of the proportions need not be provided.

Examples:

1. Na_2SO_4 sodium sulfate, preferred to disodium sulfate
2. $CaCl_2$ calcium chloride, preferred to calcium dichloride

The prefix mono- is always omitted unless its presence is necessary to avoid confusion.

Examples:

3. N_2O dinitrogen oxide
4. NO_2 nitrogen dioxide
5. N_2O_4 dinitrogen tetraoxide
6. S_2Cl_2 disulfur dichloride
7. Fe_3O_4 triiron tetraoxide
8. U_3O_8 triuranium octaoxide
9. MnO_2 manganese dioxide
10. CO carbon monoxide
11. $FeCl_2$ iron dichloride
12. $FeCl_3$ iron trichloride
13. Na_2CO_3 sodium carbonate, or disodium trioxocarbonate
14. $POCl_3$ phosphoryl trichloride
15. TlI_3 thallium triiodide
16. $Cr_{23}C_6$ tricosachromium hexacarbide
17. $K_4[Fe(CN)_6]$ tetrapotassium hexacyanoferrate
18. Na_2HPO_4 disodium hydrogenphosphate, or
                      disodium hydrogentetraoxophosphate

The use of these numerical prefixes does not affect the order of citation, which depends upon the initial letters of the names of the constituents.

However, when the name of the constituent itself starts with a multiplicative prefix (as in disulfate, dichromate, triphosphate, and tetraborate), two successive multiplicative prefixes may be necessary. When this happens, and in other cases simply to avoid confusion, the alternative multiplicative prefixes bis-, tris-, tetrakis-, pentakis-, etc. are used (see Table III) and the name of the group acted upon by the alternative prefix is placed in parentheses.

Examples:

19. $Ca(NO_3)_2$ calcium nitrate, or calcium bis(trioxonitrate)
20. $(UO_2)_2SO_4$ diuranyl sulfate, or bis(dioxouranium) tetraoxosulfate
21. $Ba[BrF_4]_2$ barium bis(tetrafluorobromate)
22. $Tl(I_3)_3$ thallium tris(triiodide)
23. $U(S_2O_7)_2$ uranium bis(disulfate)

24. $Ca_3(PO_4)_2$ tricalcium bis(phosphate) (Note 5i)
25. $Ca(HCO_3)_2$ calcium bis(hydrogencarbonate)

As a general rule, however, the primary set of multiplicative prefixes mono-, di-, tri-, tetra-, ... , poly- should always be used unless there is a risk of confusion and misunderstanding. Some miscellaneous examples are provided below.

Examples:

26. PCl_5 phosphorus pentachloride
27. $[Cr(CO)_6]$ hexacarbonylchromium
28. $Al_2(CO_3)_3$ dialuminium tricarbonate
29. $Na_4P_2O_6$ tetrasodium hexaoxodiphosphate(*P—P*)
30. $CaNa(NO_2)_3$ calcium sodium trinitrite
31. $Ca(BO_2)_2$ calcium dimetaborate
32. Na_2MnO_4 disodium manganate
33. $Zn(MnO_4)_2$ zinc dipermanganate
34. $Mg(ClO_2)_2$ magnesium dichlorite
35. $Fe_2(Cr_2O_7)_3$ diiron tris(dichromate)

I-5.5.2   Use of oxidation and charge numbers

I-5.5.2.1   *Definitions*

In some compounds, particularly those in which the oxidation state of one or more of the constituent atoms is not obvious, it may be necessary to provide more information if the precise proportions of the constituents are to be deduced. This information may be provided by the use of one of two devices, the oxidation (Stock) number, which designates oxidation state, and the charge (Ewens-Bassett) number, which designates ionic charge. The oxidation number of an element in any chemical entity is the number of charges which would remain on a given atom if the pairs of electrons in each bond to that atom were assigned to the more electronegative member of the bond pair. Neutral ligands are formally removed in their closed-shell configurations.

I-5.5.2.2   *Oxidation number*

In the Stock system, the oxidation number (see Chapter I-10) of an element is indicated by a roman numeral placed in parentheses immediately following the name (modified if necessary by an appropriate ending) of the element to which it refers. The oxidation number may be positive, negative, or zero. Zero, not a roman numeral, is represented by the usual cipher, 0. The positive sign is never used. An oxidation number is always positive unless the minus sign is explicitly used. Note that it cannot be non-integral. Non-integral numbers may seem appropriate in some cases where a charge is spread over more than one atom, but such a use is not encouraged (see Section I-4.4.1).

Note 5i. Compare $Ca_3(PO_4)_2$ with $Ca_2P_2O_7$, calcium diphosphate.

Examples:

1. UO_2^{2+} uranyl(VI)
2. PCl_5 phosphorus(V) chloride
3. PO_4^{3-} phosphate(V)
4. Na^- natride(−I)
5. $[Fe(CO)_5]$ pentacarbonyliron(0)

There are several conventions observed for inferring oxidation-numbers, whose use is particularly common in names of compounds of transition elements. Hydrogen is considered positive (oxidation number I) in combination with non-metallic elements, and negative (oxidation number −I) in combination with metallic elements. Organic radicals combined with metal atoms are treated as anions (for example, a methyl ligand is considered to be a methanide ion, CH_3^-) and the nitrosyl group (NO) is always considered to be neutral. Bonds between atoms of the same species make no contribution to oxidation number.

Examples:

6. N_2O nitrogen(I) oxide
7. NO_2 nitrogen(IV) oxide
8. Fe_3O_4 iron(II) diiron(III) oxide
9. MnO_2 manganese(IV) oxide
10. CO carbon(II) oxide
11. $FeSO_4$ iron(II) sulfate
12. $Fe_2(SO_4)_3$ iron(III) sulfate
13. SF_6 sulfur(VI) fluoride
14. Hg_2Cl_2 dimercury(I) chloride
15. $NaTl(NO_3)_2$ sodium thallium(I) nitrate
16. $(UO_2)_2SO_4$ uranyl(V) sulfate, or dioxouranium(V) sulfate
17. UO_2SO_4 uranyl(VI) sulfate, or dioxouranium(VI) sulfate
18. $K_4[Fe(CN)_6]$ potassium hexacyanoferrate(II)
19. $K_4[Ni(CN)_4]$ potassium tetracyanonickelate(0)
20. $Na_2[Fe(CO)_4]$ sodium tetracarbonylferrate(−II)
21. $[Co(NH_3)_6]ClSO_4$ hexaamminecobalt(III) chloride sulfate
22. $Fe_4[Fe(CN)_6]_3$ iron(III) hexacyanoferrate(II)

I-5.5.2.3   *Charge number*

A charge number is a number in parentheses written without a space immediately after the name of an ion, and whose magnitude is the ionic charge. Thus the number may refer to cations or anions, but never to neutral species. The charge is written in arabic numerals and followed by the sign of the charge. This system should be used only to designate the charges of ions. Statement of ionic charge often gives an unequivocal indication of stoichiometry, but multiplicative prefixes may also be used if necessary or desirable. The system is of most use for compounds of elements which exhibit a range of oxidation states. Note that, unlike in superscript charge designation, unity (1) is always indicated.

Examples:

23.	$FeSO_4$	iron(2+) sulfate (cf., Example 11)
24.	$Fe_2(SO_4)_3$	iron(3+) sulfate (cf., Example 12)
25.	$NaTl(NO_3)_2$	sodium thallium(1+) nitrate (cf., Example 15)
26.	$(UO_2)_2SO_4$	uranyl(1+) sulfate, or dioxouranium(1+) sulfate (cf., Example 16)
27.	UO_2SO_4	uranyl(2+) sulfate, or dioxouranium(2+) sulfate (cf., Example 17)
28.	$K_4[Fe(CN)_6]$	potassium hexacyanoferrate(4−) (cf., Example 18)
29.	$K_4[Ni(CN)_4]$	potassium tetracyanonickelate(4−) (cf., Example 19)
30.	$Na_2[Fe(CO)_4]$	sodium tetracarbonylferrate(2−) (cf., Example 20)
31.	$[Co(NH_3)_6]ClSO_4$	hexaamminecobalt(3+) chloride sulfate (cf., Example 21)
32.	$Fe_4[Fe(CN)_6]_3$	iron(3+) hexacyanoferrate(4−) (cf., Example 22)

I-5.6   ADDITION COMPOUNDS

The term addition compounds covers donor–acceptor complexes and a variety of lattice compounds. The methods described here are particularly relevant to compounds of uncertain structure which cannot be named according to the methods of Chapter I-10. The ending -ate has often been used in names of such adducts, particularly solvates. Since this ending is restricted to names of anions, the use in the context of addition compounds is discouraged.

The name of an addition compound is formed by connecting the names of individual compounds by long dashes, and indicating the proportions of each species after the name by an arabic number separated by a solidus from the numbers referring to other species. The whole numerical symbol is placed in parentheses and separated from the compound name by a space. The sequence of names of the individual species is the sequence in formulae (see Section I-4.6.8).

In addition compounds, the name of H_2O is water. The term hydrate has a specific meaning, a compound containing water of crystallization bound in an unspecified way.

Examples:

1.	$3CdSO_4 \cdot 8H_2O$	cadmium sulfate—water (3/8)
2.	$Na_2CO_3 \cdot 10H_2O$	sodium carbonate—water (1/10) or sodium carbonate decahydrate
3.	$Al_2(SO_4)_3 \cdot K_2SO_4 \cdot 24H_2O$	aluminium sulfate—potassium sulfate—water (1/1/24)
4.	$CaCl_2 \cdot 8NH_3$	calcium chloride—ammonia (1/8)
5.	$AlCl_3 \cdot 4C_2H_5OH$	aluminium chloride—ethanol (1/4)
6.	$2CH_3OH \cdot BF_3$	methanol—boron trifluoride (2/1)
7.	$BiCl_3 \cdot 3PCl_5$	bismuth(III) chloride—phosphorus(V) chloride (1/3)
8.	$BF_3 \cdot 2H_2O$	boron trifluoride—water (1/2)
9.	$8H_2S \cdot 46H_2O$	hydrogen sulfide—water (8/46)
10.	$8Kr \cdot 46H_2O$	krypton—water (8/46)
11.	$6Br_2 \cdot 46H_2O$	dibromine—water (6/46)

12. $CHCl_3 \cdot 2H_2S \cdot 17H_2O$ chloroform—hydrogen
                                       sulfide—water (1/2/17)

13. $Co_2O_3 \cdot nH_2O$ cobalt(III) oxide—water (1/n)

I-5.7   BORON HYDRIDES

Boron nomenclature is specialized and a preliminary treatment is given in Chapter I-11. The number of boron atoms in a boron-hydride molecule is indicated by a numerical prefix but the number of hydrogen atoms is indicated by the appropriate arabic numeral in parentheses directly following the name (see Chapter I-11).

Examples:

1. B_2H_6 diborane(6)
2. $B_{20}H_{16}$ icosaborane(16)

I-5.8   FINAL REMARKS

I-5.8.1   Appropriate choice of name

In principle, simple compounds may be named according to any one of the several methods discussed in this Chapter. For example, $NaTl(NO_3)_2$ can be named as sodium thallium dinitrate, sodium thallium(I) nitrate or sodium thallium(1 +) nitrate. However, the style to be chosen is that most appropriate to the use of the name in question. Pb_3O_4 may be named as trilead tetraoxide, but dilead(II) lead(IV) oxide ($Pb^{II}_2Pb^{IV}O_4$) is more informative. Antimony dioxide describes the compound with empirical formula SbO_2 uniquely, but the apparent correspondence to antimony(IV) oxide is misleading because Sb^{III} and Sb^V are both present in SbO_2. Certain double salts are best named using multiplicative prefixes (Section I-5.5.1). For example, the name calcium(II) sodium(I) sulfate fails to distinguish between calcium disodium bis(sulfate) ($CaSO_4 \cdot Na_2SO_4$) and calcium tetrasodium tris(sulfate) ($CaSO_4 \cdot 2Na_2SO_4$). Names of coordination anions may be more precisely descriptive, but they should be used with care in the absence of firm empirical data. For example, potassium hexafluoroantimonate ($K[SbF_6]$) is more precise than the names antimony(V) potassium fluoride and antimony potassium hexafluoride, but is misleading if the presence of $[SbF_6]^-$ has not been confirmed. Similarly, titanium zinc trioxide is preferred to titanium zincate for $TiZnO_3$.

     In case of doubt, the simpler names conveying less structural information, such as the double salt and double oxide names, are preferable.

I-5.8.2   Conclusion

This Chapter has sought to provide names which convey little more than stoichiometries. For specialized applications rather more complex rules have been developed. Some are discussed in detail in other Chapters (for example, Chapter I-10, coordination compounds; Chapter I-11, boron hydrides). Procedures for other complicated systems may be found in *Pure and Applied Chemistry* for poly(oxoanions), **59**, 1529 (1987); nitrogen hydrides, **54**, 2545 (1982); and coordination polymers, **57**, 149 (1985)] or will be published elsewhere (for example, for organometallic compounds).

I-6 Solids

CONTENTS

I-6.1   INTRODUCTION

I-6.1.1   **General**

This Chapter deals with the nomenclature of solids. Fully descriptive names for solids are very often difficult to construct and only the formulae will be given, especially when detailed structural information has to be conveyed. In certain cases, such as mineral names, point defect notation, and notation for metallic phases, the reader is referred to the detailed published literature (Note 6a).

I-6.1.2   **Stoichiometric and non-stoichiometric phases**

In binary and multi-component systems, intermediate crystalline phases (stable or metastable) may occur. Thermodynamically, the composition of any such phase is variable. In some cases, the possible variation in composition is very small, as for example in the case of sodium chloride. Such phases are called stoichiometric. However, in other phases appreciable variations in composition can occur, as for example in wustite. These are called non-stoichiometric phases (Note 6b). In general, it is possible to define an ideal composition relative to which the variations occur. This composition, called the stoichiometric composition, is usually the one in which the ratio of the numbers of different atoms corresponds to the ratio of the numbers of normal crystallographic sites of different types in the ideal (ordered) crystal. This concept can be used even when the stoichiometric composition is not included in the homogeneity range of the phase. The term 'non-stoichiometric' does not mean phases with complex formulae, but those with *variable composition*.

I-6.2   NAMES OF SOLID PHASES

I-6.2.1   **General**

Names for stoichiometric phases, such as NaCl, are derived simply as in Chapter I-5, whereas formulae are derived as presented in Chapter I-4. Although NaCl in the solid state consists of an infinite network of units, $(NaCl)_\infty$, the compound is named sodium chloride and represented symbolically as NaCl.

However, for non-stoichiometric phases and solid solutions, formulae are preferable to names, since strictly logical names tend to be inconveniently cumbersome. They should be used only when unavoidable (e.g., for indexing) and should be constructed in the following style.

Examples:
  1. iron(II) sulfide (iron deficient)
  2. molybdenum dicarbide (carbon excess)

I-6.2.2   **Mineral names**

Mineralogical names should be used only to designate actual minerals and not to define chemical composition; thus, the name calcite refers to a particular mineral (contrasted

---

Note 6a. Since the 1970 edition of the Red Book, international usage and practice in solid state nomenclature have changed significantly. The nomenclature presented here therefore deviates in many aspects from that presented in the earlier versions of these recommendations.

Note 6b. In the older literature non-stoichiometric phases are often called *berthollides*. This is no longer recommended.

with other minerals of similar composition) and is not a term for the chemical compound the composition of which is properly expressed by the name calcium carbonate.

A mineral name may, however, be used to indicate the structure type. Where possible, a name that refers to a general group should replace a more specific name. For example, large numbers of minerals have been named that are all spinels, but which have widely differing atomic constituents. In this case, the generic name *spinel* type should be used rather than the more specific names chromite, magnetite, etc. The mineral name, printed in italics, should be accompanied by a representative chemical formula. This is particularly important for *zeolite* types (Note 6c).

Examples:
1. $NiFe_2O_4$(*spinel* type)
2. $BaTiO_3$(*perovskite* type)

I-6.3 CHEMICAL COMPOSITION

I-6.3.1 Approximate formulae

The formula used in any given case depends upon how much information is to be conveyed. A general notation, which can be used even when the mechanism of the variation in composition is unknown, is to put the sign $\sim$ (read as *circa*, or approximately) before the formula.

Examples:
1. $\sim FeS$
2. $\sim CuZn$

If it is desirable to give more information, one of the notations described below may be used.

I-6.3.2 Phases with variable composition

For a phase where the variation in composition is caused solely or partially by substitution, the symbols of the atoms or groups that replace each other are separated by a comma and placed together between parentheses. If possible, the formula is written so that the limits of the homogeneity range, when one or the other of the two atoms or groups is lacking, are represented. The order of citation should be alphabetical (Note 6d).

Examples:
1. (Cu,Ni) denotes the complete range of compositions from pure Cu to pure Ni.
2. K(Br,Cl) comprises the range from pure KBr to pure KCl.

Note 6c. See Chemical Nomenclature and Formulation of Compositions of Synthetic and Natural Zeolites, *Pure Appl. Chem.*, **51**, 1091 (1979).
Note 6d. When it is desired to emphasize structural information, deviations from alphabetical order of citation of the cations are allowed (Section I-4.6.1.3).

Phases for which substitution also results in vacant positions are denoted in the same way.

Examples:

3. $(Li_2,Mg)Cl_2$ denotes the solid solution intermediate in composition between LiCl and $MgCl_2$.

4. $(Al_2,Mg_3)Al_6O_{12}$ represents the solid solution intermediate in composition between $MgAl_2O_4$ ($=Mg_3Al_6O_{12}$) and Al_2O_3(*spinel* type) ($=Al_2Al_6O_{12}$).

In more complicated cases, a notation in which there are variables which define composition should always be used. The ranges of the variables can also be indicated. Thus, a phase involving substitution of atom A for B is written $A_{m+x}B_{n-x}C_p$ ($0 \leq x \leq n$). The commas and parentheses called for above are not then required.

Examples:

5. Cu_xNi_{1-x}($0 \leq x \leq 1$) is equivalent to (Cu,Ni) but such a representation is capable of conveying more information.

6. KBr_xCl_{1-x}($0 \leq x \leq 1$) is equivalent to K(Br,Cl).

7. $Li_{2-2x}Mg_xCl_2$ ($0 \leq x \leq 1$) is equivalent to $(Li_2,Mg)Cl_2$ but shows explicitly that one vacant cation position appears for every replacement of $2Li^+$ by Mg^{2+}.

8. $Co_{1-x}O$ indicates that there are vacant cation sites; for $x=0$ the formula corresponds to the stoichiometric composition CoO.

9. $Ca_xZr_{1-x}O_{2-x}$ indicates that Zr has been partly replaced by Ca, leaving vacant anion sites; for $x=0$ the formula corresponds to the stoichiometric composition ZrO_2.

If it is desired to show that the variable x is limited to small values only, this may be done by using δ or ε instead of x. A specific composition or composition range can be indicated by stating the actual value of the variable x (or δ, or ε). This value can be written in parentheses after the general formula. However, the value of the variable may also be introduced in the formula itself. This notation can be used both for substitutional and for interstitial solid solutions (Note 6e).

Examples:

10. $Fe_{3x}Li_{4-x}Ti_{2(1-x)}O_6$($x=0.35$), or $Fe_{1.05}Li_{3.65}Ti_{1.30}O_6$

11. $LaNi_5H_x$ ($0 < x < 6.7$)

12. $Al_4Th_8H_{15.4}$

13. $Ni_{1-\delta}O$

I-6.4   POINT DEFECT (KRÖGER–VINK) NOTATION

I-6.4.1   General

In addition to the chemical composition, information about point defects, site symmetry, and site occupancy can be given by using additional symbols. These symbols may also be used to write quasi-chemical equilibria between point defects (Note 6e).

Note 6e. See also F. A. Kröger and H. J. Vink, *Solid State Phys.*, **3**, 307 (1956).

I-6.4.2 Indication of site occupation

In a formula, the main symbols indicate the species present at a certain site, defined with respect to empty space. This will generally be the symbol of an element. If a site is vacant this is denoted by the italicized symbol V (Note 6f). The site and its occupancy in a structure of ideal composition are represented by right lower indexes. The first index indicates the type of site, and the second index, separated from the first by a comma, indicates the number of atoms on this site. Thus, an atom A on a site normally occupied by A in the ideal structure is expressed by A_A; an atom A on a site normally occupied by B is expressed A_B, and $M_{M,1-x}N_{M,x}M_{N,x}N_{N,1-x}$ stands for a disordered alloy, where the ideal composition is $M_M N_N$ with all M atoms on one type of crystallographic site and all N atoms on a second type of crystallographic site. An alternative description is $(M_{1-x}N_x)_M(M_xN_{1-x})_N$. A species occupying an interstitial site (i.e., a site which is unoccupied in the ideal structure) is indicated by the subscript i.

Examples:

1. $Mg_{Mg,2-x}Sn_{Mg,x}Mg_{Sn,x}Sn_{Sn,1-x}$ shows that in Mg_2Sn some of the Mg atoms are present on Sn sites and *vice versa*.
2. $(Bi_{2-x}Te_x)_{Bi}(Bi_xTe_{3-x})_{Te}$
3. $Al_{Al,1}Pd_{Al,x}Pd_{Pd,1-x}V_{Pd,2x}$ shows that in AlPd, $(1-x)$ Pd sites are occupied by Pd, $2x$ Pd sites are vacant, x Pd atoms are on Al sites and every Al is on an Al site.
4. $Ca_{Ca,1}F_{F,2-x}V_{F,x}F_{i,x}$ shows that in CaF_2, x F sites are vacant, while x F ions are situated on interstitial sites.
5. $(Ca_{0.15}Zr_{0.85})_{Zr}(O_{1.85}V_{0.15})_O$, or $Ca_{Zr,0.15}Zr_{Zr,0.85}O_{O,1.85}V_{O,0.15}$ shows that in ZrO_2, 0.85 of the Zr sites are occupied by Zr, 0.15 of the Zr sites are occupied by Ca, and that, of the 2 oxygen sites, 1.85 sites are occupied by oxygen ions, leaving 0.15 sites vacant.
6. $C_{C,0.8}V_{C,0.2}V_{V,1}$ shows that 0.2 C-sites are vacant in a vanadium carbide.

The defect symbols can be used in writing quasi-chemical reactions.

Examples:

7. $Na_{Na} \rightarrow V_{Na} + Na(g)$ indicates the evaporation of a Na atom, leaving behind a sodium vacancy in the lattice.
8. $1/2 Cl_2(g) + V_{Cl} \rightarrow Cl_{Cl}$ indicates the incorporation of a chlorine atom on a vacant chlorine site in the lattice.

I-6.4.3 Indication of crystallographic sites

Crystallographic sites can be distinguished by subscripts, e.g., tet, oct, and dod, denoting tetrahedrally, octahedrally, and dodecahedrally coordinated sites, respectively. The use of

---

Note 6f. In certain contexts other symbols, such as a square box □, are used for vacancies, but the use of V is preferred. The element vanadium is written with the upright symbol V. However, if to avoid confusion, the element may be designated by the symbol Va (see Section I-3.3.3) or the vacancy by □.

subscripts such as a, b, . . . , which are not self-explanatory, is not approved. In some cases, such as oxides and sulfides, the number of subscripts can be reduced by defining specific symbols to indicate site symmetries, e.g. ( ) for tetrahedral sites, [ ] for octahedral sites, { } for dodecahedral sites. To avoid confusion, such enclosing marks should be restricted to cases where they are not being used to express multiplication. The meaning of the symbols should be clearly stated in the text.

Examples:

1. $Mg_{tet}Al_{oct,2}O_4$ or $(Mg)[Al_2]O_4$ denotes a normal spinel.
2. $Fe_{tet}Fe_{oct}Ni_{oct}O_4$ or $(Fe)[FeNi]O_4$ denotes $NiFe_2O_4$(*inverse spinel* type) (Note 6d).

I-6.4.4   **Indication of charges**

Charges are indicated by a right upper index. When formal charges are given, the usual convention holds: one unit of positive charge is indicated by a superscript $+$, n units of positive charge by a superscript $n+$, one unit of negative charge by superscript $-$, n units of negative charge by a superscript $n-$. Thus, A^{n+} denotes n units of formal positive charge on an atom of symbol A (Section I-3.4.1). In defect chemistry, charges are defined preferably with respect to the ideal unperturbed crystal. In this case, they are called *effective charges*. One unit of positive effective charge is shown by a superscript dot˙ (not to be confused with the radical dot as described in Section I-4.4.3) and one unit of negative effective charge by a prime ', n units of effective charge are indicated by superscript$^{n˙}$ or$^{n'}$. The use of double dots˙˙ or double primes″ in the case of two effective charges is also allowed. Thus, $A^{2˙}$ and $A^{˙˙}$ indicate that an atom of symbol A has two units of effective positive charge. Sites that have no effective charge relative to the unperturbed lattice may be indicated explicitly by a superscript cross^x.

Examples:

1. $Li_{Li,1-2x}Mg^˙_{Li,x}V'_{Li,x}Cl_{Cl}$ and $Li^x_{Li,1-2x}Mg^˙_{Li,x}V'_{Li,x}Cl^x_{Cl}$ are equivalent expressions for a substitutional solid solution of $MgCl_2$ in LiCl.
2. $Y_{Y,1-2x}Zr^˙_{Y,2x}O''_{i,x}O_3$ and $Y^x_{Y,1-2x}Zr^˙_{Y,2x}O''_{i,x}O^x_3$ are equivalent expressions for an interstitial solid solution of ZrO_2 in Y_2O_3.
3. $Ag_{Ag,1-x}V'_{Ag,x}Ag^˙_{i,x}Cl_{Cl}$ indicates that a fraction x of the Ag^+ ions is removed from the Ag sites to an interstitial site, leaving the silver site vacant.

Formal charges may be preferred in cases where the unperturbed crystal contains an element in more than one oxidation state.

Examples:

4. $La^{2+}_{La,1-3x}La^{3+}_{La,2+2x}V_{La,x}S^{2-}_4$ $(0<x<1/3)$
5. $Cu^+_{Cu,2-x}Fe^{3+}_{Cu,x}Tl^+_{Tl}Se^{2-}_{Se,1+2x}Se^-_{Se,1-2x}$ $(0<x<1/2)$ shows that Fe^{3+} partly replaces Cu^+ in $Cu^+_2Tl^+Se^{2-}Se^-$.

Free electrons are denoted by e', free holes by h˙. As crystals are macroscopically neutral bodies, the sums of the formal charges and of the effective charges must be zero.

Table I-6.1 Examples[a] of defect notation in $M^{2+}(X^-)_2$ containing a foreign ion Q

interstitial M^{2+} ion	$M_i^{\cdot\cdot}$	M atom vacancy	V_M^x
interstitial X^- ion	X_i'	X atom vacancy	V_X^x
M^{2+} ion vacancy	V_M''	normal M^{2+} ion	M_M^x
X^- ion vacancy	$V_X^{\cdot}$	normal X^- ion	X_X^x
interstitial M atom	M_i^x	Q^{3+} ion at M^{2+} site	$Q_M^{\cdot}$
interstitial X atom	X_i^x	Q^{2+} ion at M^{2+} site	Q_M^x
interstitial M^+ ion	$M_i^{\cdot}$	Q^+ ion at M^{2+} site	Q_M'
interstitial X^- ion	X_i'	free electron	e'
M^+ ion vacancy	V_M'	free hole	$h^{\cdot}$

[a] Consider an ionic compound $M^{2+}(X^-)_2$. The formal charge on M is $2+$, the formal charge on X is $-$. If an atom X is removed, one negative unit of charge remains on the vacant X site. The vacancy is neutral with respect to the ideal MX_2 lattice and is therefore indicated by V_X or V_X^x. If the electron is also removed from this site, the resultant vacancy is effectively positive, i.e. $V_X^{\cdot}$. Similarly, removal of an M atom leaves V_M, removal of an M^+ ion leaves V_M', removal of an M^{2+} ion leaves V_M''. If an impurity with a formal charge of three positive units Q^{3+}, is substituted on the M^{2+} site, its effective charge is one positive unit. Therefore it is indicated by $Q_M^{\cdot}$.

I-6.4.5 Defect clusters and use of quasi-chemical equations

Pairs or more complicated clusters of defects can be present in a solid. Such a defect cluster is indicated between parentheses. The effective charge of the cluster is indicated as an upper right index.

Examples:
  1. $(Ca_K^{\cdot} V_K')^x$ denotes a neutral defect pair in a solid solution of $CaCl_2$ and KCl.
  2. $(V_{Pb}'' V_{Cl}^{\cdot})'$ or $(V_{Pb} V_{Cl})'$ indicates a charged vacancy pair in $PbCl_2$.

Quasi-chemical reactions may be written for the formation of these defect clusters.

Examples:
  3. $Cr_{Mg}^{\cdot} + V_{Mg}'' \rightarrow (Cr_{Mg} V_{Mg})'$ describes the association reaction of a Cr^{3+} impurity in MgO with magnesium vacancies.
  4. $2Cr_{Mg}^{\cdot} + V_{Mg}'' \rightarrow (Cr_{Mg} V_{Mg} Cr_{Mg})^x$ gives another possible association reaction in the system of Example 3.
  5. $Gd_{Ca}^{\cdot} + F_i' \rightarrow (Gd_{Ca} F_i)^x$ describes the formation of a dipole between a Gd^{3+} impurity and a fluorine interstitial in CaF_2.

I-6.5 PHASE NOMENCLATURE

I-6.5.1 Introduction

The use of the Pearson notation (see also Section I-3.7.3) is recommended for the designation of the structures of metals and solid solutions in binary and more complex systems. The use of Greek letters, which do not convey the necessary information, and of the *Strukturbericht* designations, which are not self-explanatory, is not approved.

I-6.5.2   **Recommended notation**

The Pearson symbol consists of three parts: first, a lower-case italic letter (a, m, o, t, h, c) designating the crystal system; second, an italic capital letter (P, S, F, I, R) designating the lattice setting and, finally, a number designating the number of atoms in the conventional unit cell (Note 6g). Table I-3.1 summarizes the system.

Examples:

1. Cu, symbol ($cF4$), indicates copper of cubic symmetry, with face-centred lattice, containing 4 atoms per unit cell.
2. NaCl, symbol ($cF8$), indicates a cubic face-centred lattice with 8 atoms per unit cell.
3. CuS($hP12$), indicates a hexagonal primitive lattice with 12 atoms per unit cell.

If required, the Pearson symbol can be followed by the space group and a prototype formula.

Example:

4. $Ag_{1.5}CaMg_{0.5}$($hP12$, $P6_3/mmc$)($MgZn_2$ type).

I-6.6   NON-STOICHIOMETRIC PHASES

I-6.6.1   **Introduction**

There are a number of special problems of nomenclature for non-stoichiometric phases which have arisen with the improvements in the precision with which the structures of these materials have been determined. Thus, we now find reference to homologous series, non-commensurate and semi-commensurate structures, Vernier structures, crystallographic shear phases, Wadsley defects, chemical twinned phases, infinitely adaptive phases, and modulated structures. Many of the phases that fall into these classes have no observable composition ranges although they have complex structures and formulae; an example is $Mo_{17}O_{47}$. These phases, despite their complex formulae, are essentially stoichiometric and possession of a complex formula must not be taken as an indication of a non-stoichiometric compound (cf., Section I-6.1.2).

I-6.6.2   **Modulated structures**

Modulated structures possess two or more periodicities in the same direction of space. If the ratio of these periodicities is a rational number, the structures are called *commensurate*; if the ratio is irrational, the structures are called *non-commensurate*. Commensurately modulated structures exist in many stoichiometric and non-stoichiometric compounds; they may be regarded as superstructures and be described by the usual rules. Non-commensurately modulated structures are known to occur in several stoichiometric compounds (and some elements), usually in a limited temperature range; some examples are: U, SiO_2, TaS_2, $NbSe_3$, $NaNO_2$, Na_2CO_3, and Rb_2ZnBr_4.

Note 6g. Following the recommendations of IUCr, the letter C formerly used in the monoclinic and orthorhombic Bravais lattice type symbols has been replaced by S (side-face-centred). See *Acta Cryst.*, **A41**, 278 (1985).

Many modulated structures can be regarded as being composed of two or more substructures. The substructure with the shortest periodicity often represents a simple *basic structure*, while the other periodicities cause modulations of the basic structure. The basic structure often remains unchanged within a certain composition range, while the other substructures take up the change in stoichiometry. If this change takes place continuously, a non-stoichiometric phase with a non-commensurate structure results. If the change occurs discontinuously, a series of (essentially stoichiometric) *homologous compounds* with commensurate structures (superstructures of the basic structure) may result or—in the intermediate case—a series of compounds with *semi-commensurate* or *Vernier* structures.

Examples:

1. Mn_nSi_{2n-m}. The structure is of the *TiSi₂* type which has two atom substructures, the Mn array being identical to that of the Ti array in $TiSi_2$ and the Si_2 array being identical to that of the Si_2 array in $TiSi_2$. Removal of Si leads to a composition Mn_nSi_{2n-m} in which the Mn array is completely unchanged. The Si atoms are arranged in rows and, as the Si content falls, the Si atoms in the rows spread out. In this case there will be a Vernier relationship between the Si atom rows and the static Mn positions which will change as the composition varies, giving rise to incommensurate structures.

2. $YF_{2+x}O$. The structure is of the *fluorite* type with extra sheets of atoms inserted into the parent YX_2 structure. When these are ordered, a homologous series of phases result. When they are disordered, we have a non-commensurate, non-stoichiometric phase, while partial ordering will give a Vernier or semi-commensurate effect. Other layer structures can be treated in the same way.

Misfit structures consist of two or more different, often mutually incommensurate, units which are held together by electrostatic or other forces; no basic structure can be defined. The composition of compounds with misfit structures is determined by the ratio of the periodicities of their structural units and by electroneutrality.

Examples:

3. $Sr_{1-p}Cr_2S_{4-p}$ with $p = 0.29$, where chains of compositions Sr_3CrS_3 and $Sr_{3-x}S$ lie in tunnels of a framework of composition $Cr_{21}S_{36}$; the three units are mutually non-commensurate.

4. $LaCrS_3$, which is built from non-commensurate sheets of $(LaS)^+$ and $(CrS_2)^-$.

I-6.6.3   **Crystallographic shear structures**

Crystallographic shear planes (*CS* planes) are planar faults in a crystal that separate two parts of the crystal which are displaced with respect to each other. The vector describing the displacement is called the crystallographic shear vector (*CS* vector). Each *CS* plane causes the composition of the crystal to change by a small increment, because the sequence of crystal planes that produces the crystal matrix is changed at the *CS* plane. (From this it follows that the *CS* vector must be at an angle to the *CS* plane. If it were parallel to the plane, the succession of crystal planes would not be altered and no composition change would result. A planar boundary where the displacement vector is parallel to the plane is more properly called an *antiphase boundary*).

Because each *CS* plane changes the composition of the crystal slightly, the overall composition of a crystal containing a population of *CS* planes will depend upon the number of *CS* planes present and their orientation. If the *CS* planes are disordered, the crystals will be non-stoichiometric, the stoichiometric variation being due to the *CS* plane 'defect'. If the *CS* planes are ordered into a parallel array, a stoichiometric phase with a complex formula results. In this case, a change in the separation of the *CS* planes in the ordered array will produce a new phase with a new composition. The series of phases produced by changes in the spacing between *CS* planes is called an *homologous* series. The formula of the series will depend upon the type of *CS* plane in the array and the separation between the *CS* planes. A change in the *CS* plane may change the formula of the homologous series.

Examples:

1. Ti_nO_{2n-1}. The parent structure is TiO_2 (*rutile* type). The *CS* planes are the (121) planes. Ordered arrays of *CS* planes can exist, producing an homologous series of oxides with formulae Ti_4O_7, Ti_5O_9, Ti_6O_{11}, Ti_7O_{13}, Ti_8O_{15}, Ti_9O_{17}. The series formula is Ti_nO_{2n-1}, with n between 4 and 9.

2. $(Mo,W)_nO_{3n-1}$. The parent structure is WO_3. The *CS* planes are the (102) planes. Ordered arrays of *CS* planes can form, producing oxides with formulae Mo_8O_{23}, Mo_9O_{26}, $(Mo,W)_{10}O_{29}$, $(Mo,W)_{11}O_{32}$, $(Mo,W)_{12}O_{35}$, $(Mo,W)_{13}O_{38}$, and $(Mo,W)_{14}O_{41}$. The series formula is $(Mo,W)_nO_{3n-1}$, with n between 8 and 14.

3. W_nO_{3n-2}. The parent structure is WO_3. The *CS* planes are the (103) planes. Ordered arrays of *CS* planes can form, producing oxides with formulae W_nO_{3n-2}, with n between approximately 16 and 25.

I-6.6.4   Unit cell twinning or chemical twinning

This is a structure-building component in which two constituent parts of the structure are twin-related across the interface. The twin plane changes the composition of the host crystal by a definite amount (which may be zero). Ordered, closely spaced arrays of twin planes will lead to homologous series of phases. Disordered twin planes will lead to non-stoichiometric phases in which the twin planes serve as the defects. There is a close parallel between chemical twinning and crystallographic shear.

Example:

1. $(Bi,Pb)_nS_{n-4}$. The parent structure is PbS which has the *cF8* (*NaCl* type) structure. The twin planes are (311) with respect to the PbS unit cell. Two members of the homologous series are known, $Bi_8Pb_{24}S_{36}$ and $Bi_8Pb_{12}S_{24}$, but other members are found in the quaternary Ag–Bi–Pb–S system. The difference between compounds lies in the separation of the twin planes and each structure is built up of slabs of PbS of varying thickness, alternate slabs being twinned across (311) with respect to the parent structure.

I-6.6.5 Infinitely adaptive structures

In some systems it would appear that any prepared compositions can yield a fully ordered crystal structure over certain temperature and composition ranges. As the composition changes, so the structure changes to meet this need. The term *infinitely adaptive structures* has been applied to this group of substances (Note 6h) and is recommended.

Examples:
1. Compounds in the Cr_2O_3–TiO_2 system between the composition ranges $(Cr,Ti)O_{2.93}$ and $(Cr,Ti)O_{2.90}$.
2. Compounds in the Nb_2O_5–WO_3 system with block-type structure between the composition limits Nb_2O_5 and $8WO_3 \cdot 9Nb_2O_5$ ($Nb_{18}W_8O_{69}$).

I-6.6.6 Intercalates

There are a number of materials in which a host matrix has a guest species inserted into it. The process is called intercalation, and the product is called an *intercalation compound*. The noun 'intercalate' generally refers to the guest species. Common examples of intercalated materials are found in the clay silicates, layered dichalcogenides, and graphite. Intercalated materials can be designated by conventional chemical formulae such as Li_xTaS_2 ($0 < x < 1$) or by host–guest designations, such as $TaS_2:xLi$ ($0 < x < 1$). If the stoichiometry is definite, ordinary compound designations may be used as, for example, $TaS_2(N_2H_4)_{4/3}$, $TiSe_2(C_5H_5N)_{1/2}$, and KC_8 (Note 6i).

For transition metal dichalcogenides, it is sometimes necessary to distinguish between insertion of a guest element between the layers and substitution of one metal atom for another in the host structure. The former would be represented by Fe_xTiSe_2, and the latter by $Fe_xTi_{1-x}Se_2$.

I-6.7 POLYMORPHISM

I-6.7.1 Introduction

A number of chemical compounds and elements change their crystal structure with external conditions such as temperature and pressure. These various structures are termed polymorphic forms of the compounds, and in the past have been designated using a number of labelling systems, including Greek letters and roman numerals. As no consistent system has evolved, these trivial labels are discouraged. A rational system based upon crystal structure should be used wherever possible (cf., Sections I-3.7.3 and I-4.2.4).

Polytypes can be regarded as a special form of polymorphism and are not treated here.

I-6.7.2 Use of crystal systems

Polymorphs are indicated by adding an italicized symbol denoting the crystal system after the name or formula. The symbols to be used are given in Table I-3.1. For example,

Note 6h. See J. S. Anderson, *J. Chem. Soc., Dalton Trans.*, 1107 (1973).
Note 6i. In this particular instance the use of fractional indexes is permitted.

zinc sulfide (c) or ZnS(c) corresponds to the zinc blende structure or sphalerite, and ZnS(h) to the wurtzite structure. Slightly distorted lattices may be indicated by use of the circa sign $\sim$. Thus a slightly distorted cubic lattice would be expressed as ($\sim c$). In order to give more information, simple well-known structures should be designated by giving the type compound in parentheses whenever it is possible. For example, AuCd above 343 K should be designated AuCd($CsCl$ type) rather than AuCd(c).

Properties which are strongly dependent on lattice and point symmetries may require the addition of the space group to the crystal system abbreviation.

I-6.8   AMORPHOUS SYSTEMS AND GLASSES

The term *amorphous* includes subsets, such as *vitreous, glassy, non-crystalline*, and *supercooled liquid*. The main requirement is absence of long-range translational order. The clearest way to indicate amorphous nature is to use the designation (am) after the chemical formula, e.g., SiO_2(am) for amorphous silica. The terms *vitreous* and *glassy* imply short-range order (generally involving near neighbours only), connectivity (the short-range ordered units are connected in a disordered way to other similar units), and relatively high density comparable to an ordered structure. On heating to particular temperatures, glasses exhibit a second-order transition, 'the glass transition'. A vitreous or glassy material can be indicated by the designation (vit), e.g., SiO_2(vit) for silica glass (see note 6j).

No accepted system has yet been developed to express as a formula the doping of an amorphous material with another 'impurity' element. To be consistent with the previous sections a symbolism such as Si(am)H_x may be used to designate amorphous silicon doped with hydrogen. When the amount of dopant is known, it can be specified in the formula, e.g., Si(am)$H_{0.005}$.

I-6.9   FINAL REMARKS

The nomenclature in the present Chapter has been adapted to international usage in solid state chemistry. In cases such as polymorphs the nomenclature is still under discussion in the scientific community and no firm recommendation can yet be given. For complicated topics, such as graphite intercalation and polytypes, detailed nomenclatures still need to be worked out.

Note 6j. These abbreviations form part of the more extensive notation for states given in *Pure Appl. Chem.*, **54**, 1239 (1982).

I-7   Neutral Molecular Compounds

CONTENTS

I-7.1   INTRODUCTION

This Chapter is restricted to consideration of neutral molecules in which bonding is of the two centre covalent type. Acids and their direct derivatives are treated more fully in Chapter I-9 and detailed coverage of coordination compounds is given in Chapter I-10, which also deals with some organometallic compounds. Neutral boron cluster compounds are discussed in Chapter I-11.

A choice should be made, according to circumstances, from the various methods available for naming molecular compounds. For simple molecules a stoichiometric name suffices for most purposes (see Chapter I-5). For example, octacarbonyldicobalt and tetraphosphorus decaoxide are unique identifications, but are uninformative as to structural detail. To provide this, two other systems of nomenclature are recommended to construct more descriptive names.

(i) *Substitutive nomenclature*—a system commonly used for organic compounds, in which names are based on that of a parent hydride which has an ending characteristic of the class. In organic hydrides this may be -ane, -ene, -yne, or -ine but for hydrides of elements other than carbon the ending is generally -ane (the examples of -ine endings cited in the footnote 'a' to Table I-7.2, Section I-7.2.2.1 are not recommended for naming substituted derivatives). In the absence of other designators such as λ, the -ane ending is taken to convey implicitly that central atoms exhibit their usual bonding numbers (e.g., 3 for P; 4 for Si) and that all non-skeletal valencies are satisfied by an appropriate but unstated number of neutral hydrogen atoms (see Section I-7.2).

(ii) *Coordination nomenclature*—an additive system originally intended for Werner-type complexes, which makes none of the implicit assumptions of substitutive nomenclature but, by contrast, is based on the concept of connectivity. Thus coordination names describe the connective situation of every central and ligating atom in the molecule under consideration, including any hydrogen atoms (see Section I-7.3).

The use of both nomenclature systems on the same chemical structures is illustrated by the Examples below.

Examples:

		Substitutive name	Coordination name
1.	$Te(OCOCH_3)_2$	diacetoxytellane	bis(acetato)tellurium
2.	$SiCl_3(OCH_2CH_2CH_3)$	trichloro(propoxy)silane	trichloro(propan-1-olato)-silicon
3.	$Si(OCH_2CH_3)_4$	tetraethoxysilane	tetrakis(ethanolato)silicon
4.	$P(CF_3)(PHCF_3)_2$	1,2,3-tris(trifluoromethyl)-triphosphane	trifluoromethyl-bis(trifluoromethyl-phosphanido)phosphorus
5.	$Sb(C_6H_5)_2Cl$	chlorodiphenylstibane	chlorodiphenylantimony
6.	IF_5	pentafluoro-λ^5-iodane	pentafluoroiodine

I-7.2   SUBSTITUTIVE NOMENCLATURE

I-7.2.1   Introduction

This is a method of naming, commonly used for organic compounds, in which names are based on that of an individual parent hydride, usually ending in -ane, -ene or -yne. The hydride name is understood to signify a definite fixed population of hydrogen atoms attached to a skeletal structure.

Examples:

1. pentane $CH_3CH_2CH_2CH_2CH_3$
2. benzene C_6H_6
3. disilane Si_2H_6

Names for derivatives are formed by citing prefixes appropriate to the substituting monoatomic or polyatomic groups (preceded by locants when required), followed without a break by the name of the unsubstituted parent hydride. In the case of carbon and silicon, certain substituting groups may be denoted instead by appropriate suffixes.

Examples:

4. 1-bromopentane
5. nitrobenzene
6. 1,1-dichlorocyclohexane
7. hydroxystannane
8. silanol
9. cyclobutane-1,3-dithiol

These names are taken to convey that the population of hydrogen atoms in the parent hydride, e.g., pentane, stannane, silane, or cyclobutane, has been correspondingly reduced by substitution. Thus, nitrobenzene is $C_6H_5NO_2$, the NO_2 being a monovalent group replacing one hydrogen atom of benzene. Similarly, benzenepentacarboxylic acid is $C_6H(COOH)_5$.

When this method is applied to elements other than carbon, the bonding number of the skeletal atoms must either be conventionalized by definition (e.g., 4 for Si and Sn) or, if variable, indicated by appropriate designators (see Section I-7.2.2.3). These considerations apart, the method is analogous to that used in organic nomenclature and the relevant practices and conventions of the Organic Rules are followed (Note 7a).

Thus, for example, silane is SiH_4 and trisilane is $SiH_3SiH_2SiH_3$. The names for hydrides such as these in which any or all of the hydrogen atoms may be replaced by other atoms or groups are considered first.

I-7.2.2   **Hydride names**

I-7.2.2.1   *Names of mononuclear hydrides*

Substitutive nomenclature is usually confined to the following central elements: B, C, Si, Ge, Sn, Pb, N, P, As, Sb, Bi, O, S, Se, Te, Po, as illustrated by the enclosed zone shown on a section of the Periodic Table in Table I-7.1, but it may be extended to certain halogen derivatives, especially those of iodine.

The names of the mononuclear hydrides used in this nomenclature system are listed in Table I-7.2.

In the absence of any designator, the ending -ane signifies that the skeletal element exhibits its standard bonding number, namely 3 for boron (Note 7b), 4 for the Group 14

Note 7a. See the *Nomenclature of Organic Chemistry*, 1979 edition, Sections A, B, and C.
Note 7b. This concept is inadequate to cover the hydrides of polyboron clusters (see Chapter I-11), although the ending -ane in used there in a somewhat analogous manner.

elements, 3 for the Group 15 elements, and 2 for the Group 16 elements. In cases where bonding numbers other than these are exhibited, they must be indicated in the hydride name by means of an appropriate superscript appended to the Greek letter λ (Note 7c), these symbols being separated from the name from Table I-7.2 by a hyphen.

Examples:
1. PH_5 λ^5-phosphane
2. SH_6 λ^6-sulfane

This use of lambda applies to the -ane names of Table I-7.2, but not to the synonyms for these names.

Table I-7.1 Parent hydride elements

Group 13	Group 14	Group 15	Group 16	Group 17	Group 18
B	C	N	O	F	Ne
Al	Si	P	S	Cl	Ar
Ga	Ge	As	Se	Br	Kr
In	Sn	Sb	Te	I	Xe
Tl	Pb	Bi	Po	At	Rn

Table I-7.2 Mononuclear parent hydrides

BH_3	borane	NH_3	azane[a]	OH_2	oxidane[a, b]
SiH_4	silane	PH_3	phosphane[a]	SH_2	sulfane[a]
GeH_4	germane[b]	AsH_3	arsane[a]	SeH_2	selane[b]
SnH_4	stannane	SbH_3	stibane[a]	TeH_2	tellane[b]
PbH_4	plumbane	BiH_3	bismuthane[b]	PoH_2	polane

[a] Phosphine, arsine, and stibine may be retained for the unsubstituted mononuclear hydrides and for use as derived ligands and in forming certain derived groups, but they are not recommended for naming substituted derivatives. The systematic names in substitutive nomenclature for ammonia, NH_3, and water, H_2O, are azane and oxidane, respectively. These names are usually not used, but are available if required. Sulfane, when unsubstituted, is usually named hydrogen sulfide. The normal formulae H_2O, H_2S, H_2Se, H_2Te, and H_2Po have been reversed in Table I-7.2 for purposes of comparison.

[b] Names based on such other forms as oxane, germanane, selenane, tellurane, and bismane cannot be used because they are used as names for saturated six-membered heteromonocyclic rings based on the Hantzsch–Widman system [*Pure Appl. Chem.*, **55**, 409 (1983)].

Other stem names, endings, and endings with prefixes have been used for mononuclear parent hydrides with a bonding number other than the normal one, for example, phosphorane, arsorane, sulfurane, and persulfurane. Although this technique may offer certain advantages of convenience, it is not generally applicable, does not have a sound basis for extension, and has generated ambiguous names, and is therefore not recommended. Accordingly, the λ-convention is preferred for naming 'non-standard' parent hydrides.

Note 7c. This is fully described in The Treatment of Variable Valency in Organic Compounds (Lambda Convention) Recommendations 1983, *Pure Appl. Chem.*, **56**, 769 (1984) (see also Section I-7.2.2.3).

The chief usefulness of the -ane names of Table I-7.2 lies in their convenience for naming substituted derivatives and their derived radical forms, as well as the ease with which the same style can be extended into the nomenclature of chains and rings. Organic derivatives of H_2S and H_2Se will usually be named as sulfides or thiols, or selenides or selenols, according to the Organic Rules (Note 7d), whereas those of NH_3 will usually be named as amines, amides, nitriles, etc., as in the Organic Rules (Note 7e).

Examples:

3. $C_6H_5SC_6H_5$ diphenyl sulfide, or diphenylsulfane
4. CH_3CH_2SeH ethaneselenol, or ethylselane
5. $(BrCH_2)_3N$ tris(bromomethyl)amine, or tris(bromomethyl)azane

I-7.2.2.2   *Names of oligonuclear hydrides derived from elements of standard bonding number*

Names are constructed by prefixing the -ane names of the corresponding mononuclear hydride from Table I-7.2 with the appropriate multiplicative prefix (di-, tri-, tetra-, etc.) corresponding to the number of atoms of the chain bonded in series.

Examples:

1. H_2PPH_2 diphosphane
2. H_3SnSnH_3 distannane
3. $HSeSeSeH$ triselane
4. $SiH_3SiH_2SiH_2SiH_3$ tetrasilane

The names azane (for NH_3, conventionally known as ammonia but, by tradition, modified to 'amine' in one method for naming its substituted organic derivatives), diazane (for N_2H_4, commonly known as hydrazine), diazene (for $HN=NH$) (Note 7f), triazane (for NH_2NHNH_2) and so on, conform to this rule although triazane and tetraazane are unknown in the free state. However, substituted derivatives are known (see Example 5 in Section I-7.2.3.3). Names of compounds containing the $>N-N<$ unit are more fully covered by Organic Rules (Note 7g) which include the use of 'hydrazine' as a parent for naming substituted derivatives. In so far as it is applicable, that usage constitutes a partial alternative to the use of diazane (Note 7h).

The names of unsaturated compounds may be derived analogously, by extension of the appropriate Organic Rules (see also Section I-7.2.3.6).

I-7.2.2.3   *Names of oligonuclear hydrides with elements exhibiting non-standard bonding numbers* (Note 7i)

In cases where the skeletal atoms of a hydride chain are the same but one or more has a bonding number different from the values quoted in Section I-7.2.2.1, the name of the

Note 7d. See the *Nomenclature of Organic Chemistry*, 1979 edition, C-514, p. 213 and C-701, p. 247.
Note 7e. See the *Nomenclature of Organic Chemistry*, 1979 edition, C-811–C-815, p. 249, and C-821–C-843, p. 261.
Note 7f. The names 'diimide' and 'diimine' have also been used for $HN=NH$, but are not recommended: see *Pure Appl. Chem.*, **54**, 2545 (1982).
Note 7g. See the *Nomenclature of Organic Chemistry*, 1979 edition, C-921.1–C-931.5, p. 284.
Note 7h. For additional discussion of the hydrides of nitrogen, see *Pure Appl. Chem.*, **54**, 2545 (1982).
Note 7i. Hydrides of boron are dealt with in Chapter I-11.

hydride is formed as if all the atoms showed standard bonding numbers (Section I-7.2.2.2) but it is preceded by locants, one for each non-standard atom, each locant qualified without space by λ^n, where n is the appropriate bonding number. Detailed usage is shown in the Examples, and is discussed in Section D of the *Nomenclature of Organic Chemistry*, 1979 edition. The assignment of locants is as discussed below (Note 7j).

Numbering of parent hydrides with heteroatoms in non-standard valence-states follows the rules of Sections B and C of the *Nomenclature of Organic Chemistry*, 1979 edition, for numbering heteroatoms, as far as possible. When a choice is needed between the same skeletal atom in different valence-states, the one in a non-standard valence-state is preferred for assignment of the lower locant. If a further choice is needed between the same skeletal atom in two or more non-standard valence-states, preference for the lower locant is given in order of the decreasing numerical value of the bonding number, i.e. λ^6 is preferred to λ^4.

The first sentence of this statement deals with the numbering of heterogeneous chains. For homogeneous chains, the latter part of the quotation applies.

Examples:

   1 2 3   4

1. H_5SSSH_4SH $1\lambda^6, 3\lambda^6$-tetrasulfane

               (not $2\lambda^6, 4\lambda^6$)

   1 2   3   4   5

2. $HSSH_4SH_4SH_2SH$ $2\lambda^6, 3\lambda^6, 4\lambda^4$-pentasulfane

               (not $2\lambda^4, 3\lambda^6, 4\lambda^6$)

The hydrides named in this way may be hypothetical, but often substituted derivatives are known. Where all the skeletal atoms in a parent hydride exhibit the same non-standard bonding number, repetition of the λ-symbols may be avoided by means of the format shown in the alternative names of Examples 3 and 4, below.

Examples:

3. $H_4PPH_3PH_3PH_4$ $1\lambda^5, 2\lambda^5, 3\lambda^5, 4\lambda^5$-tetraphosphane,

              or $(\lambda^5)_4$-tetraphosphane

4. $HPbPbPbH$ $1\lambda^2, 2\lambda^2, 3\lambda^2$-triplumbane,

              or $(\lambda^2)_3$-triplumbane

I-7.2.3   **Names of substituted derivatives of hydrides**

I-7.2.3.1   *Use of prefixes*

The analogy with organic substitutive nomenclature is maintained. Substituents, considered as replacing hydrogen atoms, are named using prefixes in appropriate radical-form (amino, acetoxy, nitroso, etc.) and not in ligand-form (acetato). This difference cannot arise where the names of the two forms are identical. Where there is more than one kind of

Note 7j. For a fuller discussion of the lambda convention, see *Pure Appl. Chem.*, **56**, 769 (1984).

substituent, the prefixes are cited in alphabetical order before the name of the parent hydride, parentheses being used to avoid ambiguity. Multiplicative prefixes indicate the presence of two or more identical groups, and if the substituting groups are themselves substituted, the prefixes bis-, tris-, tetrakis-, etc., are used (see Table III).

I-7.2.3.2   *Substituted mononuclear hydrides*

The following names exemplify the principles expounded in Section I-7.2.3.1.

Examples:

1. $PH_2(CH_2CH_3)$ ethylphosphane
2. $Te(OCOCH_3)_2$ diacetoxytellane
3. $Sb(CH=CH_2)_3$ trivinylstibane
4. $AsBr(OCH_3)(CH_3)$ bromo(methoxy)(methyl)arsane (note 7k)
5. $Si(OCH_2CH_2CH_3)Cl_3$ trichloro(propoxy)silane (note 7k)
6. $GeH(SCH_3)_3$ tris(methylthio)germane
7. $Si(OCH_2CH_3)_4$ tetraethoxysilane

Example 7 might be regarded as a derivative of orthosilicic acid, $Si(OH)_4$, and Example 4 is an arsinous acid ester. The names of derivatives of oxoacids are considered more fully in Chapter I-9.

Where there is a choice of elements listed in Table I-7.1 to be taken as the central atom, the name is based on the parent mononuclear hydride of whichever skeletal element is convenient and appropriate. This technique is clearly not applicable to those chains which cannot sensibly be regarded as derivatives of a mononuclear hydride.

Examples:

8. $H_3CPHSiH_3$ (methylphosphanyl)silane, or methyl(silyl)phosphane, or (silylphosphanyl)methane (Note 7l)
9. $Ge(C_6H_5)Cl_2(SiCl_3)$ dichloro(phenyl)(trichlorosilyl)germane, or trichloro[dichloro(phenyl)germyl]silane

When the central atom of a hydride being considered as a parent is also one atom of a functional group as defined in organic nomenclature (e.g., the O in –OH, the C in –COOH, or the S in –SO$_2$OH), the name may be constructed using the parent of that functional group in accordance with the Organic Rules. Other groups present are then cited in the name in group prefix-form (Note 7m).

Note 7k. Parentheses are used in the manner shown in order to avoid ambiguity. Thus, in Example 4, the parentheses enclosing methyl protect that group from substitution by groups preceding it in the name.
Note 7l. Phosphanyl: H_2P–, phosphanediyl: HP<, phosphanylidene: HP=, and phosphanetriyl: –P< are all logically derived from PH$_3$, phosphane. Use of the name phosphine yields instead: phosphinyl, phosphinediyl, and phosphinetriyl, respectively. This conflicts with the widespread usage of 'phosphinyl' for the substituent group $H_2P(=O)$–, which is also currently named 'phosphinoyl' by analogy with certain organic acyl groups. Therefore, names based on phosphanes are recommended as being free from this confusion.
Note 7m. Those groups derived from parent hydrides are formed by changing the ending -ane to -anyl [or -yl for those derived from hydrides of Group 14 elements] for monovalent, -anediyl for divalent, etc., substituent groups or radicals. However, single-element radical names can be formed also by adding -io to the stem of an element-name or of its latinate form, e.g. mercurio, ferrio, zincio. Usage of this form has been taken to indicate connection without implication as to the number of bonds, although it is often evident from the context. See also Chapter I-8 for a discussion of radicals and substituent groups.

Examples:

10. $H_2As(CH_2)_4SO_2Cl$ 4-arsanylbutane-1-sulfonyl chloride

$$CH_2$$

11. $Cl_3SiSCHCHPO(OH)_2$ 2-[(trichlorosilyl)thio]cyclopropylphosphonic acid

I-7.2.3.3   *Unbranched, saturated homogeneous chains*

For the purposes of naming substituted derivatives, the chain is numbered sequentially from one end to the other and substituents are cited in radical-form in alphabetical order before the name of the appropriate hydride taken from Section I-7.2.2.1 and preceded by multiplicative prefixes as appropriate (but omitting mono-). The direction of numbering is decided by the appropriate criteria (see Section I-7.2.2.3). If there are no λ-designators involved, then the direction of numbering is determined by assigning lowest locants to the substituents, taken as a set. If, after application of these criteria in the order given, a choice remains, lowest locants are assigned to the substituent cited first in the name.

Examples:

1. $(C_2H_5)_3PbPb(C_2H_5)_3$ hexaethyldiplumbane
2. $ClSiH_2SiHClSiH_2SiH_2SiH_2Cl$ 1,2,5-trichloropentasilane
3. $(F_3C)HPP(CF_3)P(CF_3)H$ 1,2,3-tris(trifluoromethyl)triphosphane
4. $H_3GeGeGeH_2GeBr_3$ 4,4,4-tribromo-$2\lambda^2$-tetragermane

In Example 4, priority is determined by the λ-designator.

Examples:

5. $CH_3NHNHNHC_3H_7$ 1-methyl-3-propyltriazane
6. $C_3H_7SnH_2SnCl_2SnH_2Br$ 1-bromo-2,2-dichloro-3-propyltristannane

In Example 6, the locant-set is 1,2,2,3 from either end but 1-bromo is preferred to 3-bromo.

I-7.2.3.4   *Branched, saturated homogeneous chains*

The name is based on the longest available, unbranched chain, which is regarded as a parent hydride, and the names of the shorter chains which are substituent groups are appropriately cited. Once the longest chain has been chosen, it is numbered so as to give lowest locants to the substituent groups, taken as a set, at the first point of difference (Note 7n).

Note 7n. There has been some practice of naming Group 14 hydrides of general formula $E(EH_3)_4$ using the neo-prefix [as in neopentane, $C(CH_3)_4$], e.g., neopentagermane for $Ge(GeH_3)_4$, but this is not recommended.

Examples:

1.

$$\overset{1}{H_3}\overset{2}{SiSiH_2} \quad \overset{5}{SiH_2}\overset{6}{SiH_2}\overset{7}{SiH_3}$$

$$\overset{3}{HSi}\overset{4}{SiH}$$

$$H_3Si \qquad SiH_2SiH_3$$

4-disilanyl-3-silylheptasilane

2. $CH_3Pb[Pb(CH_3)_3]_3$ 1,1,1,2,3,3,3-heptamethyl-2-(trimethyl-
                                 plumbyl)triplumbane

If a choice of principal chain still cannot be made, that one is preferred as a basis for the name which bears the greatest number of other substituent groups [see the *Nomenclature of Organic Chemistry*, 1979 edition, C-13.11(h)-(k), p. 99].

Example:

3.

$$ClH_2Si \qquad\qquad SiH_2Cl$$

$$\overset{2}{Si}H\overset{3}{Si}H_2\overset{4}{Si}H$$

$$\overset{1}{Cl_3}Si \qquad\qquad \overset{5}{Si}HCl_2$$

1,1,1,5,5-pentachloro-2,4-bis(chloro-
silyl)pentasilane

I-7.2.3.5   *Chains of repeating units*

Chains consisting of alternations of 2 elements ab as in ab(ab)$_n$a, can be named by successive citation of the following name parts:

(i) a numerical prefix (Table III) denoting the number of atoms of the chain-element (a or b) cited later in Table IV.

(ii) a replacement-name term ending in 'a' which denotes component elements of the chain, corresponding to a and b cited in that order, eliding the 'a' before a or o (see Table VI).

(iii) the ending -ne.

The element occurring later in the usual sequence of Table IV is considered as terminating both ends of the chain and any other attached units are cited before the items (i), (ii), and (iii) in group-prefix form, preceded by appropriate locants. For this purpose, the chain is numbered sequentially from end to end, as in Section I-7.2.2.3. Where the chain bears substituents and alternative numberings would seem possible, the starting point and direction of numbering are chosen so as to give lowest locants to the substituent-set at the first point of difference between the two alternatives.

Examples:

1. $H_3SnOSnH_2OSnH_2OSnH_3$ tetrastannoxane
2. $HSnCl_2OSnH_2OSnH_2OSnH_2Cl$ 1,1,7-trichlorotetrastannoxane
3. $H_3SiGeH_2SiH_2GeH_3$ 1-silyldigermasilane
4. $H_3SiSSiH_2SSiH_2SSiH_3$ tetrasilathiane
5. $(CH_3)SiH_2SSi(CH_3)HSSiH_2SSiH_2(CH_3)$ 1,3,7-trimethyltetrasilathiane
6. $H_2NPHNHPHNHPHNH_2$ 1,5-diaminotriphosphazane

Rules covering more complex repetitions are under consideration.

I-7.2.3.6   *Chains containing unsaturation*

Such compounds are accommodated in this substitutive method of nomenclature by the methods used with organic alkenes and alkynes, that is, the name of the corresponding saturated-chain hydride is modified by replacing the -ane suffix with -ene in the case of a double bond and -yne in the case of a triple bond. If there is one of each, the suffix becomes -ene . . . -yne with appropriate locants; -adiene is used when there are two double bonds, and so on. In each case the position of unsaturation is indicated by means of a numerical locant immediately preceding the suffix. For more complicated cases, such as branched systems with unsaturation, the criteria for chain-seniority of the appropriate Organic Rule is to be followed, in so far as it can be taken to apply (Note 7o).

Examples:

1. $H_2NNHN=NNH_2$ pentaaz-2-ene
2. $(C_6H_5)NHN=NN=NNH(C_6H_5)$ 1,6-diphenylhexaaza-2,4-diene
3. $CH_3NHN=NCH_3$ 1,3-dimethyltriazene

Note that in Example 3 there is no need for an unsaturation-locant.

Examples:

4. $CH_3N=CHN=CH_2$ tricarbaza-1,3-diene
5. $H_2PNHPHNHP=NPH_2$ tetraphosphaz-2-ene

I-7.2.3.7   *Inhomogeneous chains*

These will be treated in more detail elsewhere but when carbon atoms are present the methods of organic replacement nomenclature may be used (see the *Nomenclature of Organic Chemistry*, 1979 edition, C-61, p. 123). In this method, the chain is named as if it were composed entirely of carbon atoms and the terminal atoms must actually be carbon, but any heteroatoms lying between them are designated by appropriate replacement terms from Table VI cited in the order given there, each preceded by its appropriate locant. These are assigned by numbering from that end which gives lower locants to the heteroatom-set and, if these are equal, from that end which gives the lowest locant to the replacement term first cited. If there is still a choice, lower locants are assigned to the sites of unsaturation. Heteroatoms joined outside the terminal carbon atoms are denoted by prefixes cited at the start of the name.

   If the chain bears any functional groups (see the *Nomenclature of Organic Chemistry*, 1979 edition, Rule C-10.3, p. 87), it is numbered from the end, giving lower numbers to principal groups considered as a set, such groups being cited as suffixes in the name ending. The naming of chains lacking two terminal carbon atoms will be described in a later publication.

Examples:

        11 10 9 8   7 6   5   4 3   2 1

1. $H_3SiNHCH_2ONHCH_2SSiH_2NHOSiH_2OCH_3$

    11-(silylamino)-2,4,10-trioxa-7-thia-5,9-diaza,3,6-disilaundecane

    (the locant sequence 2,3,4,5 etc., . . . , for the Table VI terms being preferred to the alternative 2,3,5,6, etc., . . .).

Note 7o. See the *Nomenclature of Organic Chemistry*, 1979 edition, C-13.11, p. 97.

2. $NCH{=}CHCH_2OCH_2CH{=}CH_2$
$\|$
$NCH{=}CHCH_2SCH_2CH{=}CH_2$

4-oxa-13-thia-8,9-diazahexadeca-1,6,8,10,15-pentaene

(4-oxa is preferred to 13-oxa, although the locant sequence of Table VI terms is 4,8,9,13 numbering from either end).

3. $HSCH{=}NOCH_2SeCH_2ONHCH_3$

3,7-dioxa-5-selena-2,8-diazanon-1-ene-1-thiol

I-7.2.3.8   *Monocyclic compounds*

The name of a homogeneous ring is obtained by adding the prefix cyclo- to the name of the unbranched, unsubstituted chain containing the same number of identical atoms derived as described in Section I-7.2.3.3.

Examples:

1. 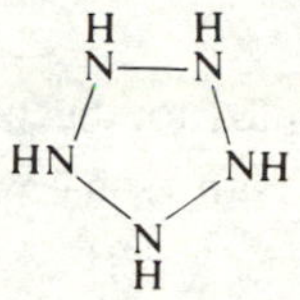   cyclopentaazane

2. 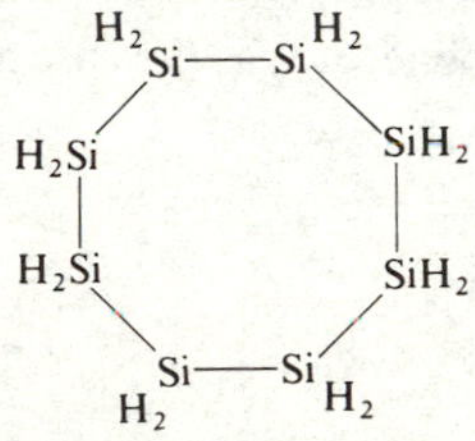   cyclooctasilane

The presence of double or triple bonds is conveyed in the name by changing the ending -ane to -ene, -yne, -adiene, -enyne, etc., as appropriate. A single multiple bond has the locant-numbering 1,2 and when there are two or more such bonds their locant-set is the lowest consistent with their position in the ring.

Examples:

3. 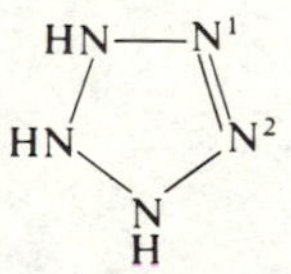   cyclopentaazene

4. 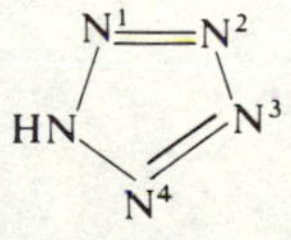   cyclopentaazadiene

For rings of repeating units of two alternating skeletal atoms the prefix cyclo- is followed by the replacement names from Table VI, cited in the reverse of the order in which they appear there.

Note that many of the monocyclic compounds under discussion here have trivial names, some of which are retained.

Examples:

	Allowed trivial name	Systematic name
5.	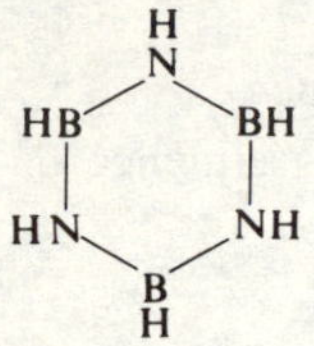 borazine	cyclotriborazane
6.	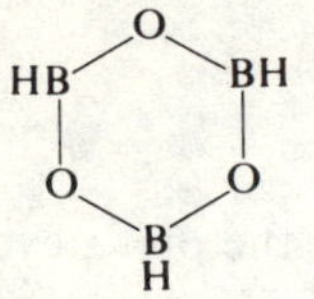 boroxin	cyclotriboroxane
7.	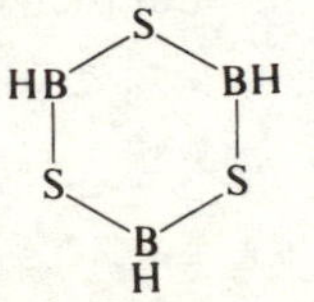 borthiin	cyclotriborathiane

The names borazole, boroxole, and borthiole should not be used for these compounds, as they imply five-membered rings in the Hantzsch–Widman system (see below).

The name ends with -ane if the repeating unit is saturated. Double bonds are denoted by changing the ending -ane to -ene, -adiene, -atriene, etc., as the case may be, preceded by the appropriate locant or locant set.

Examples:

8. 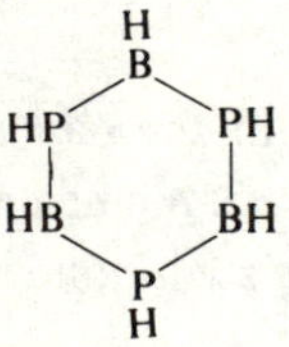   cyclotriboraphosphane

9. 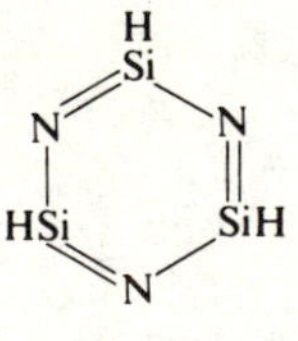   cyclotrisilaza-1,3,5-triene

10. 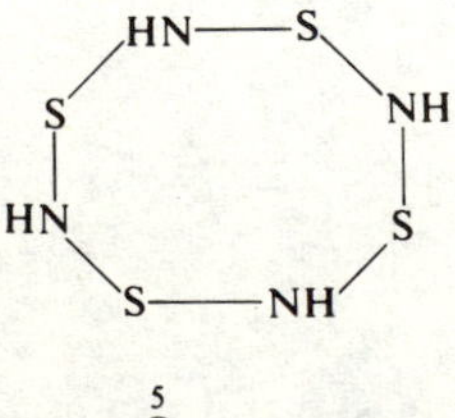   cyclotetraazathiane

11. 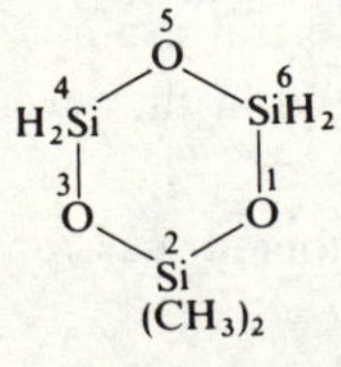   2,2-dimethylcyclotrisiloxane
(here positions 1 and 3 are equivalent)

Numbering starts at an atom cited *earliest* in Table IV and proceeds sequentially around the ring in a direction giving lowest locants to atoms occurring earliest in Table IV. In the event of a choice, lowest locants are assigned first to unsaturation and then to substituent prefixes taken as a set.

Example:

12. $(CH_3)_2 Si$ $SiHCH_2CH_3$ 4-ethyl-2,2-dimethylcyclodisilazane

Where ring-atoms display a connectivity different from their standard bonding-number (see Section I-7.2.2.1), their actual bonding-number is expressed as an arabic superscript to the Greek letter lambda following immediately after an appropriate locant.

Example:

13.   2,2,4,4,6,6-hexachloro-$2\lambda^5,4\lambda^5,6\lambda^5$-cyclotri-phosphaza-1,3,5-triene

Alternative to these methods is the Extended Hantzsch–Widman system, the procedures of which may be used also for inhomogeneous rings not capable of being named by the methods so far described. This procedure is fully presented in its most recently revised form (Recommendations 1982) in Revision of the Extended Hantzsch–Widman System of Nomenclature for Heteromonocycles, *Pure Appl. Chem.*, **55**, 409 (1983), which replaces B-1.51 and part of B-1.1 of the *Nomenclature of Organic Chemistry*, 1979 edition. Names are constructed in this class of compound by conveying the ring-size and the state of hydrogenation (saturated or unsaturated) by means of a characteristic suffix. A selection of the most important cases is given in Table I-7.3.

Table I-7.3 Suffixes used in the extended Hantzsch–Widman system

Number of atoms in ring	Unsaturated	Saturated
5	-ole	-olane (-olidine for rings containing N)
6(A)[a]	-ine	-ane
6(B)	-ine	-inane
6(C)	-inin(e)	-inan(e)
7	-epin(e)	-epan(e)
8	-ocin(e)	-ocan(e)

[a] 6(A) is for O, S, Se, Te, Bi, Hg; 6(B) is for N, Si, Ge, Sn, Pb; 6(C) is for F, Cl, Br, I, B, P, As, Sb.

Where ring-positions are occupied by diverse atoms, that cited latest in the Table IV sequence decides the suffix-form. This is preceded by appropriate replacement names cited in the order given in Table VI.

The heteroatom-set is given the lowest numerical locants consistent with sequential numbering of the ring-positions and these are cited, separated by commas and followed by a hyphen, at the start of the name. In the event of a choice, the atoms occurring earliest in Table IV are preferred for lower numbering.

Examples:

14. 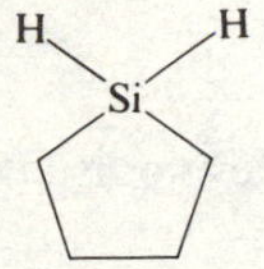 silolan(e)

15. 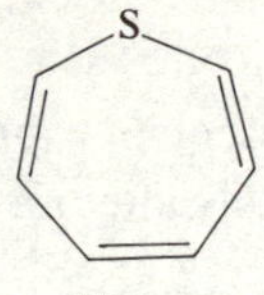 thiepin(e)

16. 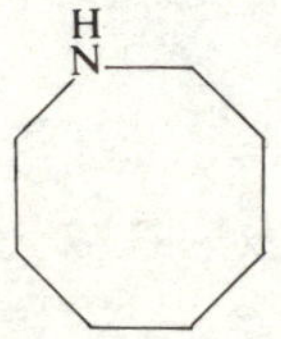 azocan(e)

17. 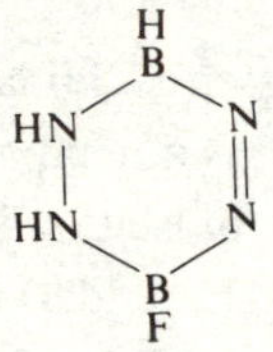 3-fluoro-1,2,3,6-tetrahydro-1,2,4,5,3,6-tetrazadiborinin(e)

18. 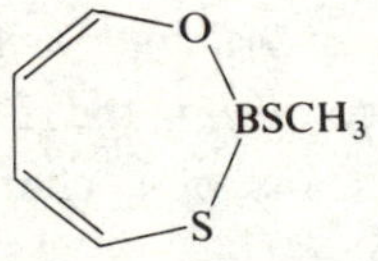 BSCH₃   2-(methylthio)-1,3,2-oxathiaborepin(e)

19. 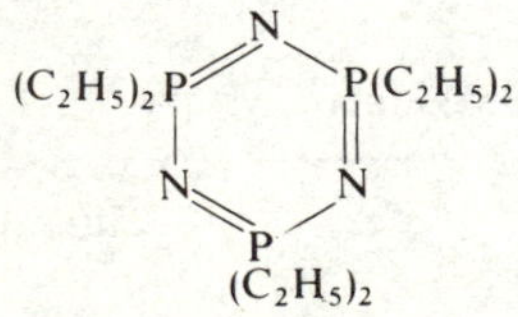 2,2,4,4,6,6-hexaethyl-1,3,5,2λ^5,4λ^5,6λ^5-triaza-triphosphinine

I-7.2.3.9   *Bi- and poly-cyclic compounds*

The methods of Section I-7.2.3.8 can be extended to other cyclic systems and these will be treated in more detail elsewhere. This brief survey is restricted to the following procedures.

   Ia Names developed by replacement nomenclature.
   Ib Names based on the citation of repeating units.
  IIa Adaption of treatment of fused rings.
  IIb The extension of the Hantzsch–Widman numbering procedure to two-ring heterocyclic systems having a fused benzene ring.

Ia *Replacement Names.* These are derived from carbocyclic-system names cited in the *Nomenclature of Organic Chemistry*, 1979 edition. When adapted for use with the compounds under discussion these trivial names are preceded by replacement prefixes (Table VI) with locants appropriate to the numbering of the carbocyclic system.

Examples:

1. 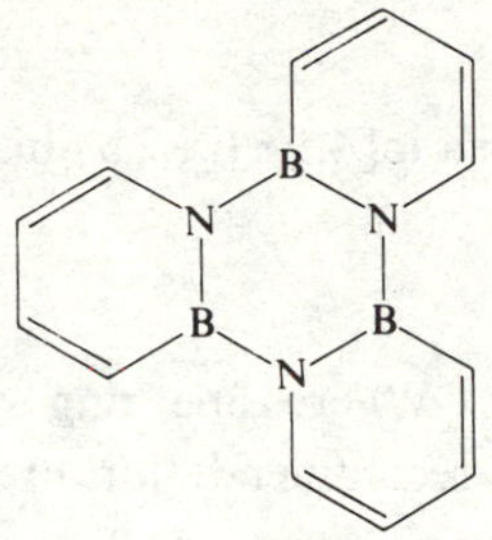 4a,8a,12a-triaza-4b,8b,12b-triboratriphenylene

2. 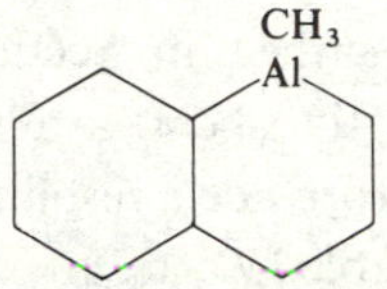 decahydro-1-methyl-1-aluminanaphthalene

3. 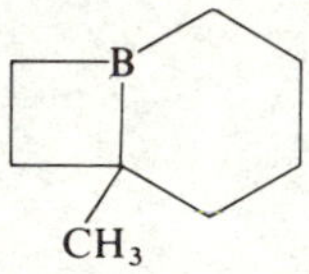 6-methyl-1-borabicyclo[4.2.0]octane

Ib *Repeating units.* The name begins with the appropriate von Baeyer term (see the *Nomenclature of Organic Chemistry*, 1979 edition, A-31 and A-32) denoting the degree of cyclicity of the molecular structure. Then follows the appropriate multiplicative prefix, and the collected replacement terms from Table VI appropriate to the repeating unit, commencing, when there is a choice, with the term cited last therein.

Example:

4. 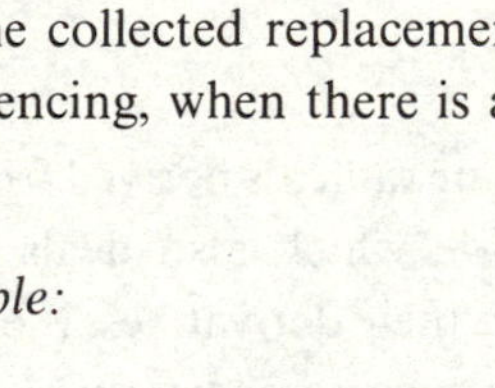 bicyclo[4.4.0]pentaborazane

This method is not generally applicable. By the more general replacement method the name for Example 4 is 1,3,5,7,9-pentaaza-2,4,6,8,10-pentaborabicyclo[4.4.0]decane or, following the procedure of Ia, decahydro-1,3,4a,6,8-pentaaza-2,4,5,7,8a-pentaboranaphthalene.

IIa *Fused rings.* The principles for naming cyclic structures considered to be composed of various fused heterocyclic organic ring-systems are set forth in detail in the *Nomenclature of Organic Chemistry*, 1979 edition, B-3. In this type of name, the components and the final fused system are all assumed to have the maximum number of non-cumulative double bonds before fusion.

Example:

5.   3*H*-[1,3]selenazolo[4,5-*c*][1,2,5]thiazaphosphinine]

IIb *Two-ring benzo-heteromonocylic systems.* Where one ring of a two-ring fused heterocyclic structure is a benzene ring, the recommendation of the *Nomenclature of Organic Chemistry*, 1979 edition, B-3.5, p. 67 may be applied. The name then begins with locants for the heteroatom-set followed by 'benzo' and then the name, formed according to the extended Hantzsch–Widman system described in Section I-7.2.3.8 for the heteromonocylic component. In the event of alternatives being available, lowest numbers are assigned to the heteroatom-set. If this does not provide a resolution, lowest numbers are assigned to the elements coming earliest in Table IV.

Examples:

6.   3-benzoxepin

7.   2,1,3-benzothiazastannole

The procedures to derive names of groups or radicals derived from any of the species described above are discussed in Chapter I-8, which also deals with cyclic ions. In numbering radicals for the purpose of naming their derivatives, the preference order for lowest locants places the position of the free valence immediately prior to unsaturation in the hierarchical series (see the *Nomenclature of Organic Chemistry*, 1979 edition, C-0.15).

I-7.3   COORDINATION NOMENCLATURE

I-7.3.1   Introduction

Although devised for molecules in which the central element is a metal atom, this nomenclature method has been extended to cover cases in which the central element is a non-metal or even an element of Group 18. The use of a name such as pentachlorophosphorus for [PCl$_5$] avoids the implication sometimes associated with a binary name, such as tungsten hexafluoride, that the compound possesses salt-like character. By contrast, where the ligand bound to the central atom is more properly considered an ion, a coordination name such as dichlorocalcium for $CaCl_2$ might appear contrived and artificial, or misleading, when compared to calcium chloride.

'Ligand' and other terms used in this section are defined in Chapter I-10.

I-7.3.2 Mononuclear coordination compounds

This section covers molecules with a single central atom. Names are formed by citing those of the ligands alphabetically before that of the central atom. Any ligands occurring more than once are collected in the name by means of appropriate multiplicative prefixes di-, tri-, tetra-, etc., for simple ligands such as chloro, benzyl, aqua, ammine, and hydroxo, and bis-, tris-, tetrakis-, etc., for complex or substituted ligands such as 2,3,4,5,6-pentachlorobenzyl and triphenylphosphine. The latter prefixes are also used to avoid any ambiguity which might attend the use of di-, tri-, etc. Those multiplicative prefixes which are not inherent parts of the ligand name do not affect the alphabetical ordering.

Examples:

1.	$[WF_5\{N(CH_3)_2\}]$	(dimethylamido)pentafluorotungsten
2.	$[GeF_4\{N(CH_3)_3\}]$	tetrafluoro(trimethylamine)germanium
3.	$[NiCl_2\{P(C_6H_5)_3\}_2]$	dichlorobis(triphenylphosphine)nickel
4.	$[B(OCH_3)_3]$	trimethoxoboron
5.	$[Ga(SO_2CH_3)_3]$	tris(methanesulfinato)gallium
6.	$[Fe(CO)_5]$	pentacarbonyliron
7.	$[Ti\{CH_2C(CH_3)_3\}_4]$	tetraneopentyltitanium
8.	$[Mn(CH_2CH=CH_2)(CO)_5]$	allylpentacarbonylmanganese

Where the position of attachment of a group could be in doubt, an unambiguous name may be achieved by means of enclosing marks, so preserving alphabetical order of ligands.

Examples:

9.	$[Hg(C_6H_5)(CHCl_2)]$	(dichloromethyl)(phenyl)mercury
10.	$[Te(CH_3)(C_5H_9)(NCO)_2]$	cyclopentyldiisocyanato(methyl)tellurium

I-7.3.3 Dinuclear coordination compounds

I-7.3.3.1 *Symmetrical dinuclear coordination compounds*

Here, each of the central atoms is of the same kind and they are identically ligated. Two methods of naming are available. (i) The following are cited in the order given, without spaces or punctuation: the appropriate multiplicative prefix for the repeated ligands, the name of the ligand, the affix di-, and finally the name of the central element. The italicized symbols for the two central elements, separated by a long dash and enclosed in parentheses, are added when it is desirable or necessary to indicate the existence of a bond between the two central element atoms.

Examples:

1.	$[(C_2H_5)_3PbPb(C_2H_5)_3]$	hexaethyldilead(*Pb—Pb*)
2.	$[(CO)_5MnMn(CO)_5]$	decacarbonyldimanganese(*Mn—Mn*)

(ii) Alternatively, the name is formed by starting with 'bis' and then citing inside parentheses the name of the half-molecule, formed by the procedure described in Section I-7.3.2.

By this method, Examples 1 and 2 above become 3 and 4, respectively.

Examples:

3. bis(triethyllead)(*Pb—Pb*)
4. bis(pentacarbonylmanganese)(*Mn—Mn*)

The long dash is the simplest of a series of designator-symbols for metal–cluster arrangements. Examples of clusters having three central atoms are briefly considered in Section I-7.3.4.3. A more detailed treatment is offered in Chapter I-10, but a satisfactory complete cluster nomenclature has yet to be developed.

I-7.3.3.2   *Non-symmetrical dinuclear coordination compounds*

These are of two types: (i) those with identical central atoms differently ligated, and (ii) those with different central atoms. In both cases names are formed by means of the procedure described in Section I-10.8.3.2 (Note 7p), which also deals with bridging groups.

Priority is assigned to the central atoms as follows. For cases of type (i) the central atom bearing the greater number of ligands is numbered 1 and the other 2. If each bears the same number of ligands, that carrying the greater number of alphabetically preferred ligands is numbered 1. For cases of type (ii) number 1 is assigned to the senior (more metallic) central element of Table IV, whatever the ligand-distribution.

In both types of compound, names are constructed by citing in turn the names of the ligand, a hyphen, the number assigned to the central atom, the Greek letter κ ('kappa') with a right superscript denoting the number of such ligands (1 being omitted for a single ligand), and the italic capital symbol for the element by which the ligand is attached to the central atom. This set constitutes the precise descriptor for the ligands and their mode of attachment. Such descriptors are cited in alphabetical order and the name ends as follows: for type (i) di-, followed by the element name of the central atom, and for type (ii) the element names of the central atoms in alphabetical order and finally their italic element-symbols, also cited in alphabetical order, enclosed in parentheses and separated by a long dash to indicate metal–metal bonding.

Examples:

$$\quad\quad\quad 2 \quad\quad\quad\quad\quad 1$$

1. $[ClGe(NHC_6H_5)_2GeCl_3]$ tetrachloro-1κ^3Cl,2κCl-bis(phenylamido-2κN)-digermanium(*Ge—Ge*)

Note 7p. This supersedes the naming procedure of Section 7.711 of the 1971 edition of the *Nomenclature of Inorganic Chemistry*, according to which one central atom and its attached ligands were expressed as a composite ligand on the other central atom. That constituted a hybrid style, appearing to apply substitutive methods to coordination nomenclature for the central atoms, but it did not provide a structurally descriptive method. The development of the 'kappa-convention' now permits the use of 'pure' coordination names with fully specific location of all attached groups and also the collection of identical ligands by means of appropriate multiplicative prefixes (see, for instance, Example 2 of this section).

     2   1

2. $[Co(CO)_4Re(CO)_5]$ nonacarbonyl-$1\kappa^5C,2\kappa^4C$-cobaltrhenium(Co—Re)

    1   2

3. $[Li\{(C_6H_5)_3Pb\}]$ triphenyl-$2\kappa^3C$-leadlithium(Li—Pb)

Where the precise positions of ligation are unknown, or are known to be mixed, the situation is appropriately met by use of names in the style of Section I-7.3.3.1.

Example:

4. $[Pb_2(C_6H_5CH_2)_2F_4]$ dibenzyltetrafluorodilead

Example 1 has positional isomers, and they could all be covered collectively, as shown by Example 5.

Example:

5. $[Ge_2(C_6H_5CH_2)(C_6H_5NH)_2Cl_3]$
    (benzyl)trichlorobis(phenylamido)digermanium

Such asymmetrical situations are treated more fully in Chapter I-10, which also deals with the further possibilities due to bridging ligands.

I-7.3.4   Other oligonuclear compounds

I-7.3.4.1   *Introduction*

Metal atoms may associate to form homonuclear and heteronuclear clusters which act as nuclear bodies in the formation of coordination entities, some of which are ionic. Such complexity, allied to the structural variations which may occur due to bridging groups, is treated in more detail in Chapter I-10. These Sections deal with a few simple molecular situations only.

I-7.3.4.2   *Chains of central atoms in coordination entities*

The procedures of Sections I-7.3.2 and I-7.3.3 may be applied to these systems but, in general, they are more conveniently treated under the procedures of Section I-7.2. Coordination names are based on that high priority atom which is nearest to the centre of the chain. Attached ligands are cited in the name according to the procedures of Section I-7.3.2.

Examples:

1. $Cl_3SiSiCl_2SiCl_3$ dichlorobis(trichlorosilyl)silicon (Note 7q),
    or octachlorotrisilicon(2 Si—Si)

2. $F(CH_3)_2SiSi(CH_3)_2Si(CH_3)_3$ (fluorodimethylsilyl)dimethyl(trimethylsilyl)silicon
    (Note 7q), or
    1-fluoroheptamethyltrisilicon(2 Si—Si)

Note 7q. The use of silyl for a group prefix name, derived from substitutive nomenclature, would not seem appropriate in a coordination name. However, its use in such circumstances is sanctified by usage. For a discussion of -io prefixes, see Section I-8.4.2.5. Application of the lambda convention may also be possible in these circumstances [see *Pure Appl. Chem.*, **56**, 769 (1984)].

In more complicated cases, the second method employed in Examples 1 and 2 gives simpler, neater names than those obtained by considering all the groups as being attached to a single central atom.

I-7.3.4.3   *Coordination compounds based on rings and clusters of central atoms*

Three like, central atoms may be arranged in a short chain, in a ring, or they may partake in tetrahedral arrangements with fourth atoms such as N, or even more complicated arrangements, a representative selection of which is considered in Chapter I-10. This Section covers one simple case only.

Example:

1. $Os(CO)_4$ dodecacarbonyltriosmium,
                                             or *cyclo*-tris(tetracarbonylosmium),
     $(CO)_4Os$———$Os(CO)_4$ or tris(tetracarbonylosmium)(3 *Os—Os*),
                                             or dodecacarbonyl-*triangulo*-triosmium

The first name gives no structural information. The other three names are applicable where the pattern of substitution is the same for each central element. The second name illustrates the general use of *cyclo* to indicate that the central atoms are joined in a ring, the third makes use of the symbolism (3 *Os—Os*) to indicate the central arrangement of osmium atoms contains three Os–Os bonds (Note 7r).

I-7.4   FINAL REMARKS

Any of the compound-types discussed above may in principle give rise to ions or radicals. These may exist free, or may be combined, as a group of atoms attached to a central structure for the purposes of substitutive nomenclature. This aspect is discussed in Chapters I-8 and I-10, but the forms chosen in constructing a name should avoid mixing the substitutive and coordination nomenclature systems. Thus, ligand-names such as disulfido, nitrito, and arsenato should not appear in hydride-based names ending in -ane, and radical names such as phosphanyl, acetyl, and boryl should not be used as anionic ligand-forms when included in a name terminated by that of a central element (coordination name). However, by established usage, ligands derived from the hydrides of elements of Group 14 are usually named in radical form, e.g., ethyl, benzyl, silyl, germyl.

Methods for naming longer-chain systems with repeating units and ring-systems of various kinds are contained in the *Nomenclature of Organic Chemistry*, 1979 edition, D-4, D-5, D-6, and D-7, pp. 373, 382, 409, and 429, respectively, though these are provisional recommendations.

Clusters of four or more central atoms present special problems which are not treated here. Molecules containing both rings and chains will be dealt with in a later publication, and cases involving multi-centre bonding, such as found in lithium–methyl tetramer, are partially covered in Chapter I-10. Such molecular species lie outside the scope of this Chapter.

Note 7r. A full structural name for this compound is provided in Example 10, Section I-10.8.3.5.

I-8   Names for Ions, Substituent Groups and Radicals, and Salts

CONTENTS

I-8.1   INTRODUCTION

This Chapter discusses the naming of ions, substituent groups and radicals, and salts. Although there are systematic rules to cover all the species discussed below, some of these involve nomenclatures such as that developed in Chapter I-10 for coordination compounds. There are extensive archaic and traditional nomenclatures, especially for oxoanions of nitrogen, sulfur, and phosphorus. Consequently, the recommendations in this Chapter contain a mixture of more systematic and traditional nomenclatures. There is also a variety of systematic methods which are complementary and sometimes give rise to alternative formulations. For example, ionic charge can be indicated by the charge number (Chapter I-2), or else the oxidation number can be used and this allows the charge to be inferred. In general, the latter usage is not encouraged, but alternative formulations are given in the text wherever they are appropriate.

I-8.2   CATIONS

I-8.2.1   **Definition and general comments**

A cation is a monoatomic or polyatomic species having one or more elementary charges of the proton. In a polyatomic species, the charge may be located on one of the atoms, or it can be delocalized. The charge on a cation can be indicated in names and formulae by use of the charge number or the oxidation number (see Section I-4.4).

The words 'ion' and 'cation' may be used following the name if this leads to clarification.

Example:

1. Cr^{3+}, or Cr^{III} ion, or chromium(3+) ion, or chromium(III) cation (see Notes 8a and 8b)

The precise meaning of a name will often depend on context. The hexaaquachromium(3+) ion may be considered as containing a chromium(3+) cation and may be referred to as such, ignoring the coordinated water. However, in the gas phase chromium(3+) may be the complete name of a precisely defined species. The oxidation number does not directly indicate charge.

I-8.2.2 Names of monoatomic cations

Monoatomic cations are named by adding in parentheses after the name of the element the appropriate charge number followed by the plus sign. Alternatively, one may use the oxidation number added in parentheses to the name of the element, followed by the words 'cation' or 'ion'.

Examples:

1. Na^+ sodium(1+) ion, sodium(I) cation (Note 8a)
2. Cr^{3+} chromium(3+) ion (Note 8b), chromium(III) cation
3. Cu^+ copper(1+) ion, copper(I) cation
4. Cu^{2+} copper(2+) ion, copper(II) cation
5. U^{6+} uranium(6+) ion, uranium(VI) cation
6. I^+ iodine(1+) ion, iodine(I) cation (Note 8c)
7. V^{5+} vanadium(5+) ion, vanadium(V) cation (Note 8c)
8. H^+ hydrogen(1+) ion, hydrogen(I) cation (Note 8d)

I-8.2.3 Names of polyatomic cations

I-8.2.3.1 *General*

The names of polyatomic cations are derived using the principles discussed in Section I-8.2.2 in all cases where the systematic names of the corresponding neutral polyatomic species can be derived. The derivations of these systematic names are described in Chapters I-7 and I-10. Detailed recommendations and exceptions are discussed below.

Note 8a. When there is no ambiguity about the charge on a cation, it may be omitted, e.g., aluminium ion for aluminium(3+) or sodium ion for sodium(1+)

Note 8b. Older names such as chromous for chromium(II), cupric for copper(II), and mercuric for mercury(II) are no longer recommended.

Note 8c. Two-letter symbols for iodine and vanadium are only permitted when required to avoid ambiguity, e.g., Id^I and Va^V (compare Section I-3.3.3).

Note 8d. The name hydron is a generic name for a proton, a deuteron, or a triton. It is used for any isotopic mixture of these cations. The name proton should be restricted to the cation derived solely from ^1H. The isotope ^2H gives rise to the deuteron and ^3H to the triton. No other element has approved specific names for individual isotopes (see Chapter I-3 and *Pure Appl. Chem.*, **60**, 1115 (1988)).

I-8.2.3.2   *Homopolyatomic cations*

The name for a homopolyatomic cation is obtained by adding the charge number to the name of the neutral species, as described in Section I-8.2.2. The oxidation number may also be used when needed. Note the use of parentheses, which may be of value in some circumstances (cf. Section I-5.3.4).

Examples:

1. $(O_2)^+$ dioxygen$(1+)$ ion
2. $(S_4)^{2+}$ tetrasulfur$(2+)$ ion
3. $(Hg_2)^{2+}$ dimercury$(2+)$ ion, or dimercury(I) cation
4. $(Bi_5)^{4+}$ pentabismuth$(4+)$ ion
5. $(H_3)^+$ trihydrogen$(1+)$ ion

I-8.2.3.3   *Cations obtained formally by the addition of hydrons to binary hydrides* (Note 8d)

The name of an ion derived by adding a hydron to a binary hydride can be obtained by adding the suffix -ium to the name of the parent hydride, with elision of any final 'e' of the name of the hydride. For polycations, the suffixes -diium, -triium, etc., are used without elision of any final 'e'. There are alternative formulations, however, as exemplified below.

Examples:

1. NH_4^+ azanium, or ammonium ion
2. $N_2H_5^+$ diazanium, or hydrazinium
3. $H_3O_2^+$ dioxidanium
4. $P_2H_5^+$ diphosphanium
5. $N_2H_6^{2+}$ diazanediium, or hydrazinium$(2+)$, or hydrazinediium

I-8.2.3.4   *Alternative names for cations obtained formally by the addition of hydrons to mononuclear binary hydrides*

Names for these simple cations can be derived as described above (Section I-8.2.3.3). Alternatively, they may be named by adding the ending -onium to a stem of the element name. This method is allowed for all the following cases and their substituted derivatives, but is preferred for Examples 1 and 5 (see also Section I-8.2.3.9).

The name oxonium is recommended for H_3O^+ as it occurs in, for example, $H_3O^+ClO_4^-$ (hydronium is not approved) and is reserved for this particular species. If the degree of hydration of the H^+ ion is not known, or if it is of no particular importance, the simpler terms hydron (Note 8d), or hydrogen ion may be used. Coordination nomenclature provides names for individual species such as $[H(H_2O)_2]^+$, namely diaquahydrogen$(1+)$.

Examples:

1. NH_4^+ ammonium, or azanium
2. PH_4^+ phosphonium
3. AsH_4^+ arsonium
4. SbH_4^+ stibonium

5.	H_3O^+	oxonium
6.	H_3S^+	sulfonium
7.	H_3Se^+	selenonium
8.	H_2F^+	fluoronium
9.	H_2Cl^+	chloronium
10.	H_2Br^+	bromonium
11.	H_2I^+	iodonium

I-8.2.3.5 *Cations obtained formally by the addition of hydrons to oxoacids and organic acids*

The names of these ions can be formed using various rules. The preferred method of nomenclature for a given case depends upon the kind of acid to which the hydron is added. There are three general methods, as detailed below.

(a) The addition of the suffix -ium to the name of the acid; this is implied in the *Nomenclature of Organic Chemistry*, 1979 edition, Rule C-82, and should be used, if at all, for organic acids only. The procedure may cause problems in languages other than English, and the derived name is not necessarily specific.

(b) The consideration of the acid as a substitution product of the oxonium ion.

(c) The use of coordination nomenclature (see Chapter I-9).

Examples:

1. $(CH_3CO_2H_2)^+$ ethanoic acidium, or acetic acidium (method a), ethanoyloxonium, or acetyloxonium (method b) (Note 8e)
2. $(H_2NO_3)^+$ dihydroxooxonitrogen cation (method c)
3. $(H_4PO_4)^+$ tetrahydroxophosphorus(v) cation (c)
4. $(H_4SO_4)^{2+}$ tetrahydroxosulfur(vi) cation (c)
5. $(CH_3CO_2H_2)(ClO_4)$ ethanoic acidium perchlorate (a), acetyloxonium perchlorate (b)
6. $(C_6H_5CH_2)(CH_3)P(NH)(OH_2)^+$

  P-benzyl-P-methylphosphinimidic acidium (a), or aquabenzyl(imido)methylphosphorus($1+$) (c)

Note the use of parentheses in these formulae; the use is optional but may be of value for longer formulae.

I-8.2.3.6 *Cations obtained formally by the addition of hydrons to various organic molecules*

A cation resulting from the formal addition of a hydron to a neutral organic compound is named by adding the suffix -ium, -diium, etc., to the parent name with elision of any final 'e' before -ium, but not before -diium, -triium, etc. Locants are added when required. For details see Rule C-82 of the *Nomenclature of Organic Chemistry*, 1979 edition.

Note 8e. The alternative pairs of names, (method a) and (method b), are not necessarily equivalent. Ethanoic acidium does not specify the site of hydronation, whereas ethanoyloxonium does.

Examples:
1. $C_5H_5NH^+$ pyridinium
2. $(CH_3)_2COH^+$ acetonium, or propan-2-ylideneoxonium

3.   imidazolium

I-8.2.3.7   *Cations produced by the formal loss of a hydride ion from a neutral molecule*

A cation produced by formal loss of a hydride ion from a neutral compound can be named by adding the suffix -ylium to the parent name, with elision of any final 'e' (see Rule C-83 of the *Nomenclature of Organic Chemistry*, 1979 edition). In the case of neutral, saturated acyclic, or monocyclic hydrocarbon names, and mononuclear derivatives of silane, germane, stannane, plumbane, and borane, -ylium replaces the ending -ane. Cations derived by formal loss of a hydroxide ion from an organic acid are named by adding the suffix -ium to the name of the corresponding acyl radical. Any necessary locant is placed immediately preceding the suffix. An acceptable name may also be derivable by citing the word cation after the name of the appropriate radical.

Examples:
1. CH_3^+ methyl cation, or methylium (n.b., carbenium is not recommended for naming CH_3^+ in inorganic contexts, due to possible confusion)
2. $CH_3CH_2^+$ ethyl cation, or ethylium
3. $CH_3C=O^+$ acetyl cation, or acetylium, or l-oxoethylium
4. PH_2^+ phosphanylium
5. SiH_3^+ silylium
6. $Si_2H_5^+$ disilanylium
7. BH_2^+ borylium

I-8.2.3.8   *Coordination cations*

The names of complex cations are derived most simply by using the coordination cation names (see Chapters I-7 and I-10). This is preferred whenever ambiguity might result.

Examples:
1. $[ICl_2]^+$ dichloroiodine(1+)
2. $[VF_4]^+$ tetrafluorovanadium(1+), or tetrafluorovanadium(v) cation
3. $[Al(POCl_3)_6]^{3+}$ hexakis(phosphoryl trichloride)aluminium(3+)
4. $[Al(H_2O)_6]^{3+}$ hexaaquaaluminium cation
5. $[CoCl(NH_3)_5]^{2+}$ pentaamminechlorocobalt(2+) ion
6. $[H(C_5H_5N)_2]^+$ bis(pyridine)hydrogen(1+) ion
7. $[H(H_2O)_2]^+$ diaquahydrogen(1+) ion
8. $[BH_2(C_5H_5N)_2]^+$ dihydrobis(pyridine)boron(1+)

I-8.2.3.9   *Substituted cations*

Names of substituted derivatives of cations can be formed from the parent names by using substitutive prefixes. Where a central atom can be designated, coordination nomenclature may be preferred. In the Examples given below, both systems are used.

Examples:

1.	$[N(CH_3)_4]^+$	tetramethylammonium ion, or tetramethylammonium cation
2.	$[PCl_4]^+$	tetrachlorophosphonium ion, or tetrachlorophosphorus $(1+)$, or tetrachlorophosphanium ion
3.	$[NF_4]^+$	tetrafluoroammonium ion
4.	$[CH_3NC_5H_5]^+$	1-methylpyridinium
5.	$[S(CH_3)(C_2H_5)(C_6H_5)]^+$	ethyl(methyl)phenylsulfonium
6.	$[SCl_3]^+$	trichlorosulfonium, or trichlorosulfanium, or trichlorosulfur$(1+)$
7.	$[NH_3(OH)]^+$	hydroxyammonium, or hydroxylaminium ion
8.	$[CH_3OH_2]^+$	methyloxonium, or methyloxidanium (Note 8f)

I-8.2.4   **Special cases**

There are a few cases where trivial, non-systematic or semi-systematic names are still allowed. Some particular examples are shown.

Examples:

1.	NO^+	nitrosyl cation
2.	NO_2^+	nitryl cation
3.	UO_2^{2+}	uranyl$(2+)$ cation, or dioxouranium$(2+)$
4.	OH^+	hydroxylium
5.	$[HOC(NH_2)_2]^+$	uronium

I-8.3   ANIONS

I-8.3.1   **Definition and general comments**

An anion is a monoatomic or polyatomic species having one or more elementary charges of the electron. In a polyatomic species, the negative charge may be located on one of the atoms, or may be delocalized. The charge may be indicated in names and formulae by using the charge number or the oxidation number. One may use the descriptive terms 'ion' or 'anion' in names, but the ending of the name should in any case indicate the presence of a negative charge. The endings are -ide (monoatomic or homopolyatomic species), -ate (coordination nomenclature, heteropolyatomic species), and -ite (used in some trivial names).

Note 8f. The name oxidanium for H_3O^+, although systematically derived from the name oxidane for water, is not preferred to oxonium.

Examples:
1. Cl^- chloride
2. S^{2-} sulfide
3. $[CoCl_4]^{2-}$ tetrachlorocobaltate($2-$)
4. $[Fe(CO)_4]^{2-}$ tetracarbonylferrate($2-$)
5. $NO_2{}^-$ nitrite

I-8.3.2   **Names of monoatomic anions**

Monoatomic anions are named by replacing the termination of the element name by -ide (see Section I-5.3.3). In many cases, contractions or variations are employed, as exemplified below.

Examples:

1.	H^-	hydride (Note 8g)	12. I^-	iodide
2.	O^{2-}	oxide	13. Br^-	bromide
3.	$^1H^-$	protide	14. Cl^-	chloride
4.	$^2H^-$	deuteride	15. F^-	fluoride
5.	S^{2-}	sulfide	16. P^{3-}	phosphide
6.	Se^{2-}	selenide	17. As^{3-}	arsenide
7.	Te^{2-}	telluride	18. Sb^{3-}	antimonide
8.	Na^-	natride	19. C^{4-}	carbide
9.	Au^-	auride	20. Si^{4-}	silicide
10.	K^-	kalide	21. B^{3-}	boride
11.	N^{3-}	nitride	22. Ge^{4-}	germide (Note 8h)

I-8.3.3   **Names of polyatomic anions**

I-8.3.3.1   *General*

The names of polyatomic anions are detailed below. The charge can be indicated in parentheses following the name (see Chapters I-3 and I-5) or implied by the oxidation number.

I-8.3.3.2   *Homopolyatomic anions*

The name of a homopolyatomic anion is derived by adding a numerical prefix, such as di-, tri-, and tetra- (Table III), and the appropriate charge number to the name of the corresponding monoatomic anion. Care should be taken to distinguish between multiple monoatomic anions and polyatomic anions. Some allowed alternatives are also given in the following examples.

Note 8g. The name hydride should be used only for the naturally occurring isotopic mixture, or as a general term for $^1H^-$, $^2H^-$, and $^3H^-$.

Note 8h. The name germide is used rather than germanide, which specifies $GeH_3{}^-$.

Examples:

	Systematic	Alternative
1. O_2^-	dioxide(1−)	hyperoxide, or superoxide (Note 8i)
2. O_2^{2-}	dioxide(2−)	peroxide
3. O_3^-	trioxide(1−)	ozonide
4. I_3^-	triiodide(1−)	
5. C_2^{2-}	dicarbide(2−)	acetylide
6. N_3^-	trinitride(1−)	azide
7. S_2^{2-}	disulfide(2−)	
8. Sn_5^{2-}	pentastannide(2−)	
9. Pb_9^{4-}	nonaplumbide(4−)	

I-8.3.3.3   *Special cases and trivial names*

There are several anions for which trivial names used in the past are no longer recommended. Other anions have trivial names which are still acceptable. A selection follows.

Examples:

1. OH^- hydroxide (not hydroxyl)
2. HS^- hydrogensulfide(1−) (hydrosulfide not recommended in inorganic nomenclature) (Note 8j)
3. NH^{2-} imide, or azanediide
4. NH_2^- amide, or azanide
5. HO_2^- hydrogendioxide(1−) (hydroperoxide not recommended in inorganic nomenclature) (Note 8j)
6. CN^- cyanide
7. $NHOH^-$ hydroxyamide, or hydroxylamide (cf., Section I-5.3.5, Example 10)
8. $N_2H_3^-$ hydrazide, or diazanide, or hydrazinide
9. NCS^- thiocyanate (Note 8k)
10. NCO^- cyanate (Note 8k)

I-8.3.3.4   *Anions derived from neutral molecules by loss of one or more hydrons*

An anion formally obtained by removal of one or more hydrons from a hydrocarbon is named by adding -ide, -diide, etc., to the name of the parent compound, with elision of any terminal 'e' present before a vowel. Alternatively, one may add the word 'anion' to the name of the corresponding radical.

Note 8i. Although O_2^- is called superoxide in biochemical nomenclature, the Commission recommends the use of the systematic name dioxide(1−), because the prefix super- does not have the same meaning in all languages. Other common names are not recommended.

Note 8j. Use of the prefix mono-, to give monohydrogensulfide, avoids confusion with hydrogen sulfide, H_2S. Note that hydrosulfide and hydroperoxide are still allowed in organic nomenclature.

Note 8k. When coordinated in mononuclear complexes, these ions may bind through either end. This has led to the use of the names isocyanate, etc., to distinguish the donor. This usage is discouraged, and the italicized donor symbol, namely, cyanato-*O*, or cyanato-*N*, should be employed.

Examples:

1. H_3C^- methanide, or methyl anion
2. $(CH_3)_2CH^-$ propan-2-ide
3. $C_6H_5^-$ benzenide
4. $C_5H_5^-$ cyclopentadienide

The names of anions formed by loss of all hydrons from structural groups such as acid hydroxyl are formed by replacing the -ic acid, -uric acid, or -oric acid ending of the acid name by -ate.

Examples:

5. $CH_3CO_2^-$ ethanoate, or acetate
6. $C_6H_5SO_3^-$ benzenesulfonate
7. NO_3^- nitrate, or trioxonitrate(v), or trioxonitrate(1−)
8. SO_4^{2-} sulfate, or tetraoxosulfate(2−)
9. PO_4^{3-} phosphate, or tetraoxophosphate(3−)

If only some of the acid hydrons are lost from an acid, the names are formed by adding 'hydrogen', 'dihydrogen', etc., before the name to indicate the number of hydrons which are still present and which can, in principle, be ionized.

Examples:

10. HCO_3^- hydrogencarbonate(1−)
11. HSO_4^- hydrogensulfate(1−), or hydrogentetraoxosulfate(vi)
12. $H_2PO_4^-$ dihydrogenphosphate(1−) (Note 8l)

Names for anions derived by formal loss of one or more hydrons from hydroxy groups of alcohols, phenols and their chalcogen analogues (characterized by suffixes such as -ol and -thiol) are formed by adding the ending -ate to the appropriate name.

Examples:

13. CH_3O^- methanolate, or methoxide (see the *Nomenclature of Organic Chemistry*, 1979 edition, Rule C-206.1). As a ligand, this ion is called 'methoxo'.
14. $C_6H_5S^-$ benzenethiolate

Names for anions derived by formal loss of one or more hydrons from non-carbon and non-oxygen atoms are formed as described for hydrocarbons, by adding the suffix -ide (or -diide, etc.) to the name of the parent compound. Alternative allowed names are in parentheses in the following Examples.

Examples:

15. NH_2^- azanide (amide)
16. HP^{2-} phosphanediide, or hydrogenphosphide(2−)

Note 8l. In the *Nomenclature of Organic Chemistry*, 1979 edition, hydrogen is always used as a separate word. However, the names used here are of coordination type, and different rules apply. In inorganic nomenclature hydrogen is regarded as a cation in the names of acids unless the name is intended to show that it is combined in an anion, as in Examples 10, 11, and 12, above.

17. SiH_3^- silanide
18. GeH_3^- germanide
19. SnH_3^- stannanide
20. $N_2H_3^-$ diazanide (hydrazide, or hydrazinide)

I-8.3.3.5   *Anions derived by adding a hydride ion to a mononuclear hydride*

Anions derived formally by the addition of a hydride ion to a mononuclear hydride are named using coordination nomenclature (see Chapters I-7 and I-10), even when the central atom is not a metal.

Examples:
1. BH_4^- tetrahydroborate(1−) (not tetrahydroboronate) (Note 8m)
2. CH_5^- pentahydridocarbonate(1−)
3. PH_6^- hexahydridophosphate(1−)

I-8.3.3.6   *Coordination nomenclature for heteropolyatomic anions*

The names of polyatomic anions which do not fall into classes mentioned above are derived from the name of the central atom using the termination -ate. Groups, including monoatomic groups, attached to the central atom are treated as ligands in coordination nomenclature. The name of the central atom, where not a metal, may be contracted.

Examples:
1. $[PF_6]^-$ hexafluorophosphate(v), or hexafluorophosphate(1−)
2. $[Zn(OH)_4]^{2-}$ tetrahydroxozincate(2−)
3. $[Sb(OH)_6]^-$ hexahydroxoantimonate(v)
4. $[SO_4]^{2-}$ tetraoxosulfate(vi), or tetraoxosulfate(2−)
5. $[HF_2]^-$ difluorohydrogenate(1−) (often named hydrogendifluoride)
6. $[BH_2Cl_2]^-$ dichlorodihydroborate(1−)

Even when the exact composition is not known, this method can be of use. The number of ligands can then be omitted, as in hydroxozincate, or zincate ion, etc.

I-8.3.3.7   *Substituted anions*

Names of substituted derivatives of anions are formed from the name of the formal parent compound by using the numbers and names of the substituents as prefixes. When a central atom can be identified, the use of coordination nomenclature (see above) may be preferred.

Examples:
1. $[B_{10}Cl_{10}]^{2-}$ decachlorodecaborate(2−)
2. $C(C_6H_5)_3^-$ triphenylmethanide

Note 8m. 'Hydro' to represent 'hydrido' or 'hydrogen' is sanctioned by usage in boron nomenclature (see Chapter I-11), but is not to be used in other contexts.

3. $(CH_3CO)_2CH^-$ 2,4-dioxopentan-3-ide
4. Cl_3C^- trichloromethanide
5. $PbCl_3^-$ trichloroplumbate$(1-)$, or trichloroplumbanide
6. CH_3CO^- 1-oxoethanide, or acetyl anion

I-8.3.3.8   *Oxoacid anions*

Although it is quite practical to treat oxygen in the same manner as ordinary ligands and use it in the naming of anions by coordination nomenclature, some names having the suffix -ite (indicating a lower-than-maximum oxidation state) are useful and therefore are still permitted.

Examples:

1. NO_2^- nitrite   4. $S_2O_4^{2-}$ dithionite
2. AsO_3^{3-} arsenite   5. ClO_2^- chlorite
3. SO_3^{2-} sulfite   6. ClO^- hypochlorite

A full list of permitted alternative names for oxoacids and derived anions can be found in Chapter I-9.

I-8.4   SUBSTITUENT GROUPS OR RADICALS

I-8.4.1   Definitions

The term radical is used here in the sense of an atom or a group of atoms having one or more unpaired electrons. An atom or group of atoms of which the name is used as a prefix in substitutive nomenclature has traditionally also been referred to as a radical. Usually such species do not have independent existence and the use of the word 'radical' in this context is discouraged in the *Compendium of Chemical Terminology, IUPAC Recommendations*, Blackwell Scientific Publications, Oxford, 1987. Radicals can be neutral (CH_3, NO), negative (O_2^-), or positive (UO_2^+). Charged radicals are discussed in Section I-8.4.3. Section I-8.4.2 deals mainly with neutral radicals, and with the names of substituent groups (Note 8n).

     An extensive selection of radical names is given in Table VIII.

I-8.4.2   Systematic names of substituent groups or radicals

I-8.4.2.1   *General names*

The names of groups which can be regarded as substituents in organic compounds or as ligands on metals are often the same as the names of the corresponding radicals. To emphasize the kind of species being described, one may add the word 'group' to the name of the species. Except for certain trivial names, names of uncharged groups or radicals usually end with -yl. Carbonyl is an allowed trivial name for the ligand CO. The

Note 8n. It should be noted that transition-metal compounds such as VCl_4, $CrCl_3$, and $CoBr_2$ may be considered as free radicals. However, they are not normally treated as such and are not so considered here.

superscript dot in the formula is used only for free radicals. Note that in the nomenclature of solid state phases (Chapter I-6) the superscript dot can indicate a charge. See also Section I-4.4.3.

Examples:
1. $(CH_3)^{\cdot}$ methanyl, or methyl
2. $(NO)^{\cdot}$ nitrosyl
3. $(NH)^{2\cdot}$ nitrene, or azanediyl

I-8.4.2.2   *Radicals and groups with special names*

Certain neutral and cationic radicals containing oxygen (or chalcogens) have, regardless of charge, special names ending in -yl. These names (or derivatives of these names) are used only to designate compounds consisting of *discrete* molecules or groups. Prefixes thio-, seleno-, and telluro- are allowed to indicate the replacement of oxygen by sulfur, selenium, and tellurium, respectively.

Examples:
1. HO   hydroxyl
2. CO   carbonyl
3. NO   nitrosyl
4. NO_2 nitryl (Note 8o)
5. PO   phosphoryl
6. SO   sulfinyl, or thionyl (Note 8p)
7. SO_2 sulfonyl, or sulfuryl (Note 8p)
8. S_2O_5 disulfuryl
9. SeO   seleninyl
10. SeO_2 selenonyl
11. HOO   hydrogenperoxyl, or perhydroxyl, or hydroperoxyl
12. CrO_2 chromyl
13. UO_2 uranyl
14. NpO_2 neptunyl
15. PuO_2 plutonyl (and similarly for other actinoids)
16. ClO   chlorosyl
17. ClO_2 chloryl
18. ClO_3 perchloryl (and similarly for other halogens)

Such names can also be used in the names of more complex molecules or in ionic species.

Note 8o. The name nitroxyl should not be used for this group because of the use of the trivial name nitroxylic acid for H_2NO_2. Although nitryl is firmly established in English, nitroyl may be a better model in other languages.

Note 8p. The former names are preferred, but the latter are allowed. The variant to be used in any particular case depends on the circumstances. Thus, sulfuryl is used in inorganic radicofunctional nomenclature and sulfonyl is used in organic substitutive nomenclature. See the *Nomenclature of Organic Chemistry*, 1979 edition, C-0.1, p. 112, and C-0.2, p. 85.

Examples:

19. $COCl_2$ carbonyl dichloride
20. $PSCl_3$ thiophosphoryl trichloride
21. S_2O_5ClF disulfuryl chloride fluoride
22. $SO_2(N_3)_2$ sulfonyl diazide, or sulfuryl diazide (Note 8p)
23. SO_2NH sulfonyl imide, or sulfuryl imide (Note 8p)
24. IO_2F iodyl fluoride
25. $NOCl$ nitrosyl chloride

I-8.4.2.3   *Radicals or groups derived formally by loss of a hydrogen atom from a molecular hydride*

Radicals or groups derived formally by the removal of one hydrogen atom from any position of a molecular hydride are named by adding the suffix -yl to the name, with the elision of any final 'e'. For radicals derived from neutral, saturated, acyclic and monocyclic hydrides, and from borane, silane, germane, stannane, and plumbane, -yl replaces the ending-ane (cf. Section I-8.4.3.2). Thus methyl and silyl are preferred to methanyl and silanyl. Although inconsistent with the systematic names derived by the present recommendations, such contractions are widely used, and therefore permitted.

Examples:

1. $(SiH_3)^{\cdot}$ silyl
2. $(Si_2H_5)^{\cdot}$ disilanyl
3. $(SnCl_3)^{\cdot}$ trichlorostannyl
4. $(PPh_4)^{\cdot}$ tetraphenylphosphoranyl,   or   tetraphenyl-λ^5-phosphanyl (Note 8q)
5. $(NH_2NH)^{\cdot}$ diazanyl, or hydrazinyl
6. $(C_6H_5NH)^{\cdot}$ phenylazanyl, or benzeneaminyl
7. $(CH_3)^{\cdot}$ methyl
8. $(HOCH_2CH_2)^{\cdot}$ 2-hydroxyethyl
9. $(HC{\equiv}CCH_2)^{\cdot}$ prop-2-yn-1-yl
10. $(BH_2)^{\cdot}$ boryl

Radicals or groups derived by loss of two or more hydrogen atoms from one or more positions of a molecular hydride (or of a substituted derivative) are named by adding the suffixes -diyl, -triyl, etc., as appropriate to the name of the parent compound but without elision of any final 'e'.

The names 'carbene', 'nitrene', and 'silene' has been used for the diradicals $(CH_2)^{2\cdot}$, $(NH)^{2\cdot}$, and $(SiH_2)^{2\cdot}$, and for substituted derivatives and even for complexes. In the present rules such names are not used to name specific diradicals. This name and related names, such as 'nitrene' $(NH)^{2\cdot}$ and 'silene' $(SiH_2)^{2\cdot}$, are not recommended. They may, however, be useful to indicate the presence of such groups in complex structures.

Examples:

11. $^{\cdot}OO^{\cdot}$ dioxidane-1,2-diyl (Note 8r)
12. $^{\cdot}HNNH^{\cdot}$ diazane-1,2-diyl (Note 8r)

Note 8q. Phosphorane and λ^5-phosphane are alternative names for PH_5 [see the *Nomenclature of Organic Chemistry*, 1979 edition, Section D, and *Pure Appl. Chem.*, **56**, 719 (1984)].
Note 8r. Not to be confused with the ground states of dioxygen, diazene, and ethene, respectively.

13. ˙H$_2$CCH$_2$˙   ethane-1,2-diyl (Note 8r)
14. (CH$_2$)$^{2\cdot}$ carbene, or methanediyl, or methylene, or methylidene
15. (NH)$^{2\cdot}$ azanediyl, or azanylidene
16. (BH)$^{2\cdot}$ boranediyl, or boranylidene, or borylidene
17. (H$_2$BH$_2$B)$^{2\cdot}$ diborane(6)-1,1-diyl, or diboran(6)ylidene
18. (CH$_3$N)$^{2\cdot}$ methylnitrene, or methylazanediyl, or methylazanylidene
19. (PH)$^{2\cdot}$ phosphanediyl, or phosphanylidene (trivial name, phosphini-
             dene)
20. (H$_3$CC)$^{3\cdot}$ ethane-1,1,1-triyl, or ethylidyne

In these di- and poly-radicals, the location of the radical centres can be indicated by the positions of the dots in structural formulae. When the specific locations of the radical centres are not known, formulae such as (O$_2$)$^{2\cdot}$, (C$_2$H$_4$)$^{2\cdot}$, and (C$_2$H$_3$)$^{3\cdot}$ should be used (see Section I-4.4.3).

I-8.4.2.4   *Substituent groups or radicals derived from hydroxy-compounds and from aldehydes*

The names of radicals derived from hydroxy-compounds, such as alcohols, acids, and phenols or from hydroperoxides are constructed by using the name of the group substituent as a prefix to 'oxyl', 'peroxyl', or 'dioxyl'.

Examples:
  1. (CH$_3$O)˙   methyloxyl (usually abbreviated to methoxyl)
  2. (C$_6$H$_5$O)˙   phenyloxyl (usually abbreviated to phenoxyl)
  3. (CH$_3$CO$_2$)˙   acetyloxyl (usually abbreviated to acetoxyl)
  4. [CH$_3$CO(OO)]˙   acetylperoxyl, or acetyldioxyl
  5. (NH$_2$CH$_2$CH$_2$O)˙   2-aminoethoxyl

Names of substituent groups derived formally from aldehydes may be based on the oxohydrocarbon name or, alternatively, acyl names are used (see the *Nomenclature of Organic Chemistry*, 1979 edition, C-403, p. 185).

Examples:
  6. (CH$_3$CO)˙   acetyl, or 1-oxoethanyl, or 1-oxoethyl
  7. (C$_6$H$_5$CO)˙   benzoyl, or phenylcarbonyl

I-8.4.2.5   *Use of the names of inorganic and organic substituent groups as prefixes*

The names of groups which can be regarded as substituents in organic compounds or as ligands on metals are often the same as the names of the corresponding radicals. This is especially true of organic radicals ending with -yl, -ylidene, etc. Radicals not derived from hydrides are named differently. For full details, see the *Nomenclature of Organic Chemistry*, 1979 edition, Sections A, B, C, and D.

Element substituent group names are formed by adding -io to a stem derived from the element names (Table VII). For Cl, Br, and I, these -io names should be used only when the valency of the radical or group is greater than 1. When the valency is 1, the forms

115

'chloro', 'bromo', and 'iodo', respectively, are used. In general, however, the use of such radical names is not intended to convey information about bond orders. In inorganic nomenclature, the use of the affix 'thio' is restricted to indicating replacement of =O by =S. The 'io' names of Table VII are to be used in substitutive nomenclature only (Note 8s), and they are not necessary in coordination nomenclature (see Chapter I-10).

Examples:

1. Cl–	chlorio	4. Ti–	titanio
2. Br–	bromio	5. Fe–	ferrio
3. S–	sulfurio	6. Er–	erbio

Polyatomic groups are not necessarily named systematically, but the -o termination is also used for their radical and substituent group names.

Examples:

7. ClS–	chlorothio	11. OCN–	cyanato-*N*
8. Cl$_2$I–	dichloroiodio	12. CH$_3$N(H)–	methylamino
9. NC–	cyano	13. NCS–	thiocyanato-*S*
10. NCO–	cyanato-*O*	14. SCN–	thiocyanato-*N*

Radicals or groups based on a central element (with or without ligand) have names which invariably take the 'io' endings of Table VII. The names are assembled according to the principles of coordination nomenclature (see Chapter I-10).

Examples:

15. Na–   sodio
16. ClHg–   chloromercurio
17. F$_5$S–   pentafluorosulfurio (Note 8t)
18. OI–   oxoiodio, or iodosyl (Note 8t)
19. (OC)$_4$Co–   tetracarbonylcobaltio

I-8.4.3   Charged radicals or substituent groups

I-8.4.3.1   *General*

When there is more than one charge centre present in a single structure, and when it is impossible to derive a name on the principles expounded for cations (Section I-8.2) and anions (Section I-8.3), names are derived by choosing one ionic centre (preferably anionic) as a basis. This centre is then used as the parent and is named according to Section I-8.2 or I-8.3. The names of the other centres in the compound are then used in prefix form.

Note 8s. In organic substitutive nomenclature, inorganic 'io' groups form one or more single bonds, each of which replaces one hydrogen atom in the parent. The endings -ylidene and -ylidyne from substitutive nomenclature indicate that the radical is partner in a double and a triple bond, respectively.

Note 8t. Alternative names for Examples 17 and 18 are pentafluoro-λ^6-sulfanyl and oxo-λ^3-iodanyl, respectively (see *Pure Appl. Chem.*, **56**, 719 (1984)).

Example:

1. $[(ClHg)C_5H_5N^+]COO^-$ [4-(chloromercurio)-1-pyridinio]formate,
   or 4-(chloromercurio)pyridinium-1-carboxylate

I-8.4.3.2   *Cationic groups*

Names of cationic radicals or groups are derived by adding suffixes such as -yl, -ylidene, and -diyl with the appropriate locants, after the -ium ending of the name of the cation derived according to Section I-8.3. Alternatively, the monovalent cation ending, -onium, may be changed to -onio. Names of monocationic centres formally derived from polynuclear structural units by loss of a hydrogen atom from the cationic site are obtained by changing the cation ending from -ium to -io.

Examples:

1. H_3N^+- ammoniumyl, or azaniumyl, or ammonio
2. $(CH_3)_2S^+-$ dimethylsulfoniumyl, or dimethylsulfaniumyl, or dimethylsulfonio
3. $CH_3CH_3^+-$ ethan-1-ium-1-yl
4. $N\equiv N^+-$ diazyn-1-ium-1-yl, or diazonio
5. $C_5H_5N^+-$ pyridin-1-ium-1-yl, or pyridinio
6. H_3As^+- arsonio, or arsaniumyl
7. HBr^+- bromonio, or bromoniumyl
8. F_3N^+- trifluoroammonio or trifluoroazaniumyl, or trifluoroammoniumyl

I-8.4.3.3   *Anionic groups*

Names of anionic radicals, formed formally by loss of hydrons from parent hydrides, are derived by adding suffixes such as -yl, -ylidene, and -diyl with the appropriate locants to the anion name obtained according to Section I-8.2, with the elision of 'e' of the -ide ending before 'y' or a vowel.

Examples:

1. H_2C^-- methanidyl
2. HN^-- azanidyl, or amidyl
3. $N^-=$ azanidylidene
4. $N^{2-}-$ azanediidyl

Names for anionic centres formally derived by loss of hydrons from all the chalcogen atoms in acids are obtained by changing the -ic acid or -ate ending to -ato. The termination -ido is a contraction of the anion ending -ide and the radical or group ending-o.

Examples:

5. $-CO(O)^-$ carboxylato
6. $-SO(O)_2^-$ sulfonato
7. $-PO_2(O)^{2-}$ phosphonato
8. O^-- oxido, or oxidanidyl

9.	S$^-$–	sulfido, or sulfanidyl
10.	(SS)$^-$–	disulfido, or disulfanidyl
11.	(OSe)$^-$–	selenidooxy, or selenidyloxy

I-8.4.4   Special cases

It should be noted that the same group may have different names in inorganic and organic nomenclature. Names of purely organic compounds, of which many are important in the chemistry of coordination compounds, should be derived in accordance with the *Nomenclature of Organic Chemistry*, 1979 edition.

I-8.5   SALTS

I-8.5.1   Definition of a salt

A salt is a chemical compound consisting of a combination of cations and anions. However, if the cation H_3O^+ is present the compound is normally described as an acid. Compounds may have both salt and acid character. When only one kind of cation and one kind of anion are present, the compound is named as a binary compound, following the guidelines given in Chapter I-5 for non-structural stoichiometric names, the ions being named according to Sections I-8.2 and I-8.3. When the compound contains more than one kind of cation and/or anion, it is still considered to be a salt, and can be named following the guidelines below.

When combining cation and anion names to form the name of a salt, the charge indications and the words 'ion', 'anion', and 'cation' should be omitted. For example, the name of common salt is written 'sodium chloride', not 'sodium cation chloride anion'. When polyatomic cations and/or anions are involved, enclosing marks should be used to avoid possible ambiguity. Thus, $Tl^I I_3$ is thallium (triiodide) and $Tl^{III}I_3$ is thallium triiodide. Alternatively, thallium(I) triiodide and thallium(III) triiodide would suffice.

I-8.5.2   Salts containing acid hydrogen

Salts containing both a hydron which is replaceable and one or more metal cations are called 'acid salts'. Names are formed by adding the word 'hydrogen', with numerical prefix where necessary, after the name of cation(s), to denote the replaceable hydrogen in the salt. 'Hydrogen' is followed without space by the name of the anion (see Note 8l). In certain cases, inorganic anions may contain hydrogen which is not easily replaceable. When it is bound to oxygen and it has the oxidation state of $+I$, it will still be denoted by 'hydrogen', though salts containing such anions cannot be designated acid salts.

Examples:

1.	$NaHCO_3$	sodium hydrogencarbonate
2.	LiH_2PO_4	lithium dihydrogenphosphate
3.	K_2HPO_4	dipotassium hydrogenphosphate
4.	$CsHSO_4$	caesium hydrogensulfate, or caesium
		hydrogentetraoxosulfate(VI),
		or caesium hydrogentetraoxosulfate(1−)

In the *Nomenclature of Organic Chemistry*, 1979 edition, hydrogen is used as a separate word (see also Sections I-8.3.3.4 and I-8.5.3.2, and Note 8l).

I-8.5.3   Double, triple, etc., salts

I-8.5.3.1   *Basic recommendations*

The names of cations generally precede the names of anions in English. In languages where cation names are placed after the anion names, the equivalents of double, triple, etc., may be added immediately after the anion name.

I-8.5.3.2   *Cations*

The names of cations other than hydrogen are cited in alphabetical order (see Chapter I-4). (This implies that differences may occur in order of citation between formulae and names, and between names in different languages.) Hydrogen is always cited last among the cations. Of course, when it is required to show that the hydrogen is part of an anion, its name is combined with the anion name, and it is cited at the appropriate position amongst the anions. When its mode of binding is unknown, a stoichiometric name should be used, treating the hydrogen as a cation. The prevalence of hydrated cations, which are in reality complex ions, would disturb the alphabetical order, but generally the hydration is ignored and the cation order is not perturbed. If it is necessary to draw attention to the presence of a particular hydrated cation, the cation name is obtained by applying the rules for complex ions, and the name takes its place accordingly in the alphabetical sequence.

Examples:
1. $KMgF_3$ — magnesium potassium fluoride
2. $NaTl(NO_3)_2$ — sodium thallium(I) nitrate, or sodium thallium dinitrate
3. $KNaCO_3$ — potassium sodium carbonate
4. $AlK(SO_4)_2 \cdot 12H_2O$ — aluminium potassium sulfate—water (1/12), or aluminium potassium bis(sulfate)—water (1/12), or hexaaquaaluminium potassium bis(sulfate) hexahydrate
5. $Na(UO_2)_3[Zn(H_2O)_6](CH_3CO_2)_9$ — hexaaquazinc sodium triuranyl nonaacetate
6. $NaNH_4HPO_4 \cdot 4H_2O$ — ammonium sodium hydrogenphosphate tetrahydrate
7. $MgNH_4PO_4 \cdot 6H_2O$ — ammonium magnesium phosphate hexahydrate

I-8.5.3.3   *Anions*

The names of anions are cited in alphabetical order. This may result in different names in different languages and also differences from the order in formulae. The stoichiometric method for indicating the proportions of constituents may be used when necessary.

Examples:

1. $NaCl \cdot NaF \cdot 2Na_2SO_4$, $Na_6ClF(SO_4)_2$ hexasodium chloride fluoride bis(sulfate)

2. $NaCl \cdot NaF \cdot 2NaHSO_4$, $Na_4ClF(HSO_4)_2$ tetrasodium chloride fluoride bis(hydrogensulfate)

3. $Ca_5F(PO_4)_3$ pentacalcium fluoride tris(phosphate)

4. $KCl \cdot 2Na_2SO_4$, $KNa_4Cl(SO_4)_2$ potassium tetrasodium chloride bis(sulfate)

The multiplicative numerical prefixes bis-, tris-, etc., are used in connection with the above anions, because di-, tri-, tetra-, etc., have been pre-empted to designate condensed anions (see Chapter I-9 and Table III).

I-8.5.4   Oxide and hydroxide salts

This section deals with the so-called 'basic' salts, which were formerly called oxy- (or oxo-) and hydroxy-salts. For the purpose of nomenclature, the salts are considered as double salts containing O^{2-} and OH^- anions. Therefore Section I-8.5.3 can be applied in its entirety.

In some languages the citation in full of all the separate anion names presents no trouble and is strongly recommended (e.g., copper chloride oxide) to the exclusion of the oxo-form wherever possible. In other languages, however, names such as 'chlorure et oxyde double de cuivre' are so far removed from current practice that the older system of using 'oxy' and 'hydroxy' (as in oxychlorure de cuivre) may be retained.

Examples:

1. $MgCl(OH)$ magnesium chloride hydroxide
2. $BiCl(O)$ bismuth chloride oxide
3. $VO(SO_4)$ vanadium(IV) oxide sulfate
4. $CuCl_2 \cdot 3Cu(OH)_2$, $Cu_2Cl(OH)_3$ dicopper chloride trihydroxide
5. $ZrCl_2(O) \cdot 8H_2O$ zirconium dichloride oxide octahydrate
6. $ZnI(OH)$ zinc hydroxide iodide
7. $CoO \cdot NiBr_2$ cobalt(II) nickel(II) dibromide oxide

I-8.5.5   Double oxides and hydroxides

There is a large variety of double, triple, etc., oxides and hydroxides. In describing these compounds the terms 'mixed oxides' and 'mixed hydroxides' are not recommended.

Many double oxides and hydroxides may be classified as belonging to one of several distinct groups, each having its own characteristic structure type, which is sometimes named after some well known mineral of the same group (such as ilmenite, perovskite, spinel, and garnet). When compounds having analogous structures are being compared, deviations from alphabetical order of the cations are allowed, as in $CaTiO_3$, $UAlO_3$, $LaGaO_3$. Names such as uranium aluminate may suggest structure and it is preferable to name such compounds as double oxides and double hydroxides, unless there is clear and generally accepted evidence that separate cations and oxo- or hydroxo-anions are present in the structure.

When a systematic name is used, the name of the structure type may be added in parentheses after the name of the compound. When the name of the structure type is added, both the formula and the name should be in accordance with the structure (see Chapter I-6).

Examples:

1. $Al_2Ca_4O_7 \cdot nH_2O$ dialuminium tetracalcium heptaoxide hydrate
2. $AlCa_2(OH)_7 \cdot nH_2O$ aluminium dicalcium heptahydroxide hydrate
3. $Ca_3[Al(OH)_6]_2$ tricalcium bis(hexahydroxoaluminate)
4. $AlLiMn^{IV}_2O_4(OH)_4$ aluminium lithium dimanganese(IV) tetrahydroxide tetraoxide
5. $MgTiO_3$ magnesium titanium trioxide (*ilmenite* type)
6. $NaNbO_3$ sodium niobium trioxide (*perovskite* type) (Note 8u)
7. $LaAlO_3$ lanthanum aluminium trioxide (*perovskite* type) (Note 8u)
8. Fe_2NiO_4 diiron(III) nickel(II) tetraoxide (*spinel* type)

I-8.6   FINAL REMARKS

In the present Chapter a detailed treatment of the naming of cations, anions, radicals, and salts has been given. The examples are as simple as possible, and in general based upon compounds treated in previous Chapters. The principles of this Chapter should be applicable to derive names for ions and radicals of more complicated species, dealt with in Chapter I-10 and in various inorganic nomenclature documents which have appeared or will appear in *Pure Appl. Chem.*

Several examples of organic species have been included in the present Chapter, as these frequently occur in inorganic compounds, for example, as ligands. The names of these ions groups and radicals are generally derived as prescribed in the *Nomenclature of Organic Chemistry*, 1979 edition. Organic practice has sometimes been modified in this Chapter, and the placing of the structural indicator has not always been consistent. Thus, names such as propan-2-onium and 2-propanonium are used, both versions would appear to be sanctioned in the Introduction to the 1969 version of the Organic Rules (see the *Nomenclature of Organic Chemistry*, 1979 edition, p. xvii).

Note 8u. Deviations from alphabetical order are allowed when compounds with analogous structures are compared.

I-9 Oxoacids and Derived Anions

CONTENTS

INTRODUCTION

Many of the compounds which have traditional names which include the word 'acid' are not acids in classical terms. Inorganic chemistry is developing in such a way that names based on function are disappearing, and nomenclature is based preferably on composition and structure, rather than on chemical properties. Chemical properties such as acidity depend on the reaction medium and a compound named as an acid might well function as a base in some circumstances. Moreover, there is no justification for distinguishing acidity for naming purposes from any one of several chemical properties, all equally important. Consequently it is preferred to give a name indicating composition and structure rather than a name which emphasizes one particular chemical property.

The nomenclature of acids has a long tradition and it would be unrealistic to systematize acid names fully and alter drastically the commonly accepted names of important and well-known substances. However, there is no reason to provide trivial names which could have a very limited use for newly prepared inorganic compounds.

The present aim is to provide a systematic way to name any compound relevant to inorganic chemistry which has the chemical property described by the term 'acid'. The preferred principles used are those employed for naming coordination compounds (Chapter I-10). These names are built to show clearly the composition and the structure. This may give rather longer names than those which will still be allowed for some very important and widely used chemicals, such as sulfuric acid and nitric acid, but the overall value of the system more than compensates for this drawback.

Thus, these recommendations will preserve the more useful of the traditional names and provide a systematic, simple, and rational way to name new compounds.

I-9.2   DEFINITION OF THE TERM OXOACID

An oxoacid is a compound which contains oxygen, at least one other element, at least one hydrogen bound to oxygen, and which produces a conjugate base by loss of positive hydrogen ion(s) (hydrons). The limits of this class of compound are dictated by usage rather than rules (Note 9a).

Oxoacids have been extensively used and studied and many of them therefore have names established by a long practice. The oldest names, such as 'oil of vitriol', are trivial. Later these names were superseded because they were found to be inconvenient and names reflecting chemical information, in this case the acid property (as, for example, with sulfuric acid), were coined. Names for the various derivatives of the parents were developed from these names. This semi-systematic approach has limitations, and has also led to ambiguities and inconsistencies.

I-9.3   FORMULAE

In a formula, the hydrogen atoms which give rise to the acid property are cited first, then comes the central atom, and finally the atoms or groups of atoms surrounding the central

Note 9a. The usage of the term oxoacid in organic chemistry is different. See the *Nomenclature of Organic Chemistry*, 1979 edition, C-4.1, p. 189.

atom. These last are cited in the following order: oxygen atoms which are bound to the central atom only, followed by other atoms and groups of atoms ordered according to coordination–nomenclature rules, that is, ionic ligands precede neutral ligands. Within each class, the order of citation is the alphabetical order of the symbols of the ligating atom.

Examples:

1. H_2SO_4
2. H_2SO_3
3. H_2SO_5
4. $H_4P_2O_6$, or $(HO)_2OPPO(OH)_2$
5. $H_4P_2O_7$, or $(HO)_2OPOPO(OH)_2$
6. HSO_3Cl

Exceptions can be made when using line formulae or semi-structural formulae (see Chapter I-4) to convey structural information, such as in polynuclear species or in organic derivatives of oxoacids (see the *Nomenclature of Organic Chemistry*, 1979 edition, Section D).

Examples:

7. $C_6H_5SO_3H$
8. $(C_6H_5)_2PO_2H$, or $(C_6H_5)_2PO(OH)$

I-9.4   TRADITIONAL NAMES

I-9.4.1   History

Some traditional names (a selection is in Table I-9.1) were introduced by Lavoisier. Under his system, oxoacids were given a two-word name, the second word being 'acid'. In the first word, the endings -ous or -ic were added to the stem of the name, intended to indicate the content of oxygen, which is known today to be related to the oxidation states of the central atom. Unfortunately, these endings do not describe the same oxidation states in different families of acids. Thus sulfurous acid and sulfuric acid refer to oxidation states IV and VI, whereas chlorous acid and chloric acid refer to oxidation states III and V.

An extension of this system became necessary as more related acids were recognized. The prefixes hypo- (for very low oxidation states) and per- (for very high oxidation states) were introduced. The prefix per- should not be confused with the syllable in the ligand name peroxo-. Finally, it became necessary to use other prefixes, ortho-, pyro-, and meta-, to distinguish acids differing in the 'content of water'.

These traditional names do not provide specific information on the number of oxygen atoms, or the number of hydrogen atoms, whether acidic or not. The use of prefixes is not always consistent; for instance, hypo- has been associated with the -ous ending (hyponitrous acid) and with the -ic ending (hypophosphoric acid). In the case of sulfur acids, two classes of acid occur, one with the stem 'sulfur' and the other with the stem 'thio'. Moreover, in substitutive nomenclature other names such as '-sulfonic acid' for

$-SO_3H$, and '-sulfinic acid' for $-SO_2H$ were developed, thereby forsaking the restriction of -ic to the higher oxidation state.

As discussed above, the important chemical property of acidity is highly solvent-dependent, but a traditional nomenclature emphasizes this property by using the word 'acid' in the name. The aim of the systematic coordination nomenclature presented here is to describe a composition and a structure, not a chemical property. Consequently, a specialized word such as 'acid' has no place in it. This is the hydrogen nomenclature discussed in Section I-9.5 below. However, in recognition of current practice, the acid nomenclature of Section I-9.6 is retained as an alternative. This is only partly systematic. Finally, the mononuclear oxoacids of phosphorus and arsenic are so numerous and have received such an immense variety of semi-systematic names, many still used, that it has been found necessary to treat them separately in Sections I-9.9 and I-9.10.

I-9.4.2   **Allowed traditional names for acids and their derived anions**

It is recommended that retained traditional names be limited to very common compounds having names established by a long practice. Systematic names should be used for all other cases. A list of these traditional names which are retained for present use is given in Table I-9.1. The use of -ous, -ic, per-, hypo-, ortho-, and meta- should be restricted to those compounds and to their derivatives; their anions are named by changing -ous into -ite and -ic into -ate. In addition, and exceptionally, sulfur and phosphorus compounds lose the syllables 'ur' and 'orus', respectively, from the acid name when it is converted to the anion name.

I-9.5   HYDROGEN NOMENCLATURE

I-9.5.1   **Name construction**

These systematic names are derived as if the acids were salts (see Chapter I-8) and the acidic hydrogen atoms are treated as if they were the cations of a salt. Coordination nomenclature is applied. The names consist of two words. The first word specifies the acidic hydrogen atoms, that is, it is the word hydrogen qualified by any necessary multiplicative prefix. The second word contains the anion name built in four parts according to coordination nomenclature (see Chapter I-10). The first part contains all the ligand names, i.e., the non-acid hydrogen atoms and the ligands attached to the central atoms; they are cited in alphabetical order; if several atoms or groups of atoms of the same kind occur, multiplicative prefixes are used: di-, tri-, tetra-, penta-, hexa-, etc. (see Table III). The second part is the stem of the central element name (the stems are contained in the second column of Table I-9.2). The third part is always the ending -ate attached to the element root name (Table I-9.2, column 3). The fourth part, always in parentheses, is either the formal ionic charge number or the oxidation number. These numbers are really redundant when a multiplicative prefix qualifies the number of hydrogen cations and in simple cases may be omitted. These 'hydrogen' names do not apply to oxoacids of the transition metals (see Table I-9.1).

Table I-9.1 Names for common oxoacids and their anions[a]

Formula	Traditional name	Traditional anion name	Hydrogen nomenclature	Acid nomenclature
H_3BO_3	boric acid	borate	trihydrogen trioxoborate	trioxoboric acid
$(HBO_2)_n$	metaboric acid	metaborate	poly[hydrogen dioxoborate(1−)]	polydioxoboric acid
$H_2B_2(O_2)_2(OH)_4$	perboric acid	perborate	dihydrogen tetrahydroxo-di-μ-peroxo-diborate(2−)	tetrahydroxodi-(μ-peroxo)diboric(2−) acid
H_4SiO_4	orthosilicic acid[b]	orthosilicate	tetrahydrogen tetraoxosilicate	tetraoxosilicic acid
$(H_2SiO_3)_n$	metasilicic acid	metasilicate	poly(dihydrogen trioxosilicate)	polytrioxosilicic acid
H_2CO_3	carbonic acid	carbonate	dihydrogen trioxocarbonate	trioxocarbonic acid
HOCN	cyanic acid[c]	cyanate	hydrogen nitridooxocarbonate	nitridooxocarbonic acid
HONC	fulminic acid	fulminate	hydrogen carbidooxonitrate	carbidooxonitric acid
HNO_3	nitric acid	nitrate	hydrogen trioxonitrate(1−)	trioxonitric acid
HNO_2	nitrous acid	nitrite	hydrogen dioxonitrate(1−)	dioxonitric acid
HPH_2O_2	phosphinic acid	phosphinate	hydrogen dihydridodioxophosphate(1−)	dihydridodioxophosphoric acid
H_3PO_3	phosphorous acid	phosphite	trihydrogen trioxophosphate(3−)	trihydridotrioxophosphoric(2−) acid
H_2PHO_3	phosphonic acid	phosphonate	dihydrogen hydridotrioxophosphate(2−)	hydridotrioxophosphoric(2−) acid
H_3PO_4	phosphoric acid	phosphate	trihydrogen tetraoxophosphate(3−)	tetraoxophosphoric acid
	orthophosphoric acid[b]	orthophosphate		
$H_4P_2O_7$	diphosphoric acid	diphosphate	tetrahydrogen μ-oxo-hexa-oxodiphosphate	μ-oxo-hexaoxodiphosphoric acid
$(HPO_3)_n$	metaphosphoric acid	metaphosphate	poly[hydrogen trioxophosphate(1−)]	polytrioxophosphoric acid
$(HO)_2OPPO(OH)_2$	hypophosphoric acid	hypophosphate	tetrahydrogen hexaoxodiphosphate $(P—P)(4−)$	hexaoxodiphosphoric acid

Formula	Acid name	Anion name	Systematic acid name	Systematic anion name
H_3AsO_4	arsenic acid	arsenate	tetraoxoarsenic acid	trihydrogen tetraoxoarsenate
H_3AsO_3	arsenous acid	arsenite	trioxoarsenic acid	trihydrogen trioxoarsenate(3−)
H_2SO_4	sulfuric acid	sulfate	tetraoxosulfuric acid	dihydrogen tetraoxosulfate
$H_2S_2O_7$	disulfuric acid	disulfate	μ-oxo-hexaoxodisulfuric acid	dihydrogen μ-oxo-hexaoxodisulfate
$H_2S_2O_3$	thiosulfuric acid	thiosulfate	trioxothiosulfuric acid	dihydrogen trioxothiosulfate
$H_2S_2O_6$	dithionic acid	dithionate	hexaoxodisulfuric acid	dihydrogen hexaoxodisulfate(S—S)
$H_2S_2O_4$	dithionous acid	dithionite	tetraoxodisulfuric acid	dihydrogen tetraoxodisulfate(S—S)
H_2SO_3	sulfurous acid	sulfite	trioxosulfuric acid	dihydrogen trioxosulfate
H_2CrO_4	chromic acid	chromate	tetraoxochromic acid	
$H_2Cr_2O_7$	dichromic acid	dichromate	μ-oxo-hexaoxodichromic acid	
$HClO_4$	perchloric acid	perchlorate	tetraoxochloric acid	hydrogen tetraoxochlorate
$HClO_3$	chloric acid	chlorate	trioxochloric acid	hydrogen trioxochlorate
$HClO_2$	chlorous acid	chlorite	dioxochloric acid	hydrogen dioxochlorate
$HClO$	hypochlorous acid	hypochlorite	monooxochloric acid	hydrogen monooxochlorate
HIO_4	periodic acid	periodate	tetraoxoiodic acid	hydrogen tetraoxoiodate
HIO_3	iodic acid	iodate	trioxoiodic acid	hydrogen trioxoiodate
H_5IO_6	orthoperiodic acid	orthoperiodate	hexaoxoiodic(5−) acid	pentahydrogen hexaoxoiodate(5−)
$HMnO_4$	permanganic acid	permanganate	tetraoxomanganic(1−) acid	
H_2MnO_4	manganic acid	manganate	tetraoxomanganic(2−) acid	

[a] The occurrence of a name in this list does not necessarily imply that the species designated has a discrete existence. For example, perboric acid probably has yet to be isolated, but perborates are well known. Both perboric acid and perborate are listed here for completeness and to demonstrate the methodology.

[b] For a discussion of the use of ortho-, etc., see Section I-9.4.1.

[c] Isocyanic acid is HNCO; this acid is not an oxoacid since hydrogen is not bound to an oxygen atom.

Table I-9.2 Names of anions of oxoacids

Element name	Stem	Anion name
actinium	actin-	actinate
aluminium	alumin-	aluminate
antimony	antimon-	antimonate or stibate[a]
arsenic	arsen-	arsenate
beryllium	beryll-	beryllate
bismuth	bismuth-	bismuthate
boron	bor-	borate
bromine	brom-	bromate
cadmium	cadm-	cadmate
carbon	carbon-	carbonate
cerium	cer-	cerate
chlorine	chlor-	chlorate
chromium	chrom-	chromate
cobalt	cobalt-	cobaltate
copper	cupr-	cuprate
gallium	gall-	gallate
germanium	german-	germanate
gold	aur-	aurate
hafnium	hafn-	hafnate
indium	ind-	indate
iodine	iod-	iodate
iridium	irid-	iridate
iron	ferr-	ferrate
lead	plumb-	plumbate
manganese	mangan-	manganate
mercury	mercur-	mercurate
molybdenum	molybd-	molybdate
nickel	nickel-	nickelate
niobium	niob-	niobate
nitrogen	nitr-	nitrate
osmium	osm-	osmate
palladium	pallad-	palladate
phosphorus	phosph-	phosphate
platinum	platin-	platinate
radium	rad-	radate
rhenium	rhen-	rhenate
rhodium	rhod-	rhodate
ruthenium	ruthen-	ruthenate
scandium	scand-	scandate
selenium	selen-	selenate
silicon	silic-	silicate
silver	argent-	argentate
sulfur	sulf-	sulfate
tantalum	tantal-	tantalate
thallium	thall-	thallate
technetium	technet-	technetate
tellurium	tellur-	tellurate
thorium	thor-	thorate
tin	stann-	stannate
titanium	titan-	titanate

Table I-9.2 (*Continued*)

Element name	Stem	Anion name
tungsten	tungst-	tungstate or wolframate[a]
uranium	uran-	uranate
vanadium	vanad-	vanadate
xenon	xenon-	xenonate
zinc	zinc-	zincate
zirconium	zircon-	zirconate

[a]Wolframate and stibate are allowed alternatives to tungstate and antimonate, respectively.

Several acceptable names can be generated as described above (see Example 1 below). In the Examples following it, only a single name has been selected.

Examples:

1. H_2SO_4 hydrogen tetraoxosulfate(2−), or hydrogen tetraoxosulfate(VI), or dihydrogen tetraoxosulfate
2. H_2SO_3 dihydrogen trioxosulfate
3. $HClO_4$ hydrogen tetraoxochlorate(1−)
4. $HClO_3$ hydrogen trioxochlorate(V)
5. $HClO_2$ hydrogen dioxochlorate(1−)
6. $HClO$ hydrogen monooxochlorate (Note 9b)
7. H_5IO_6 pentahydrogen hexaoxoiodate(5−)
8. H_2SO_2 dihydrogen dioxosulfate

I-9.5.2   **Derivatives obtained formally by replacement of oxygen atoms**

I-9.5.2.1   *Peroxoacids*

In these compounds the group −OO− is regarded as replacing an oxygen atom. It may be bridging or terminal, and its ligand name is 'peroxo'.

Examples:

1. HNO_4 hydrogen dioxoperoxonitrate(1−)
2. H_2SO_5 hydrogen trioxoperoxosulfate(2−)
3. H_3PO_5 hydrogen trioxoperoxophosphate(3−)

I-9.5.2.2   *Thioacids*

Acids derived from oxoacids by formal replacement of oxygen atoms by sulfur atoms are called thioacids. In the names of such compounds, sulfur is designated by 'thio'.

Note 9b. The prefix mono- is normally considered redundant, and is only used when specifically required to avoid confusion.

Examples:

1. H_2SO_3S hydrogen trioxothiosulfate(2−)
   (usually written $H_2S_2O_3$)
2. H_3AsS_3 hydrogen trithioarsenate(3−)
3. H_3AsS_4 hydrogen tetrathioarsenate(3−)
4. H_2CS_3 dihydrogen trithiocarbonate (Note 9c)

I-9.5.2.3   *Replacement of oxygen by groups other than peroxo and sulfur*

Acids obtained by full formal replacement of all oxygen atoms are named according to the principles of Chapter I-10.

Examples:

1. $H[PF_6]$ hydrogen hexafluorophosphate(1−)
2. $H[AuCl_4]$ hydrogen tetrachloroaurate(1−)
3. $H_2[PtCl_4]$ dihydrogen tetrachloroplatinate(2−)
4. $H_4[Fe(CN)_6]$ tetrahydrogen hexacyanoferrate(4−)
5. $H[B(C_6H_5)_4]$ hydrogen tetraphenylborate(1−)

I-9.5.3   **Functional derivatives of acids**

I-9.5.3.1   *Principles of nomenclature employed*

These derivatives are obtained by formal replacement of an OH group by another group. Names are derived according to coordination nomenclature.

I-9.5.3.2   *Acid halogenides (halides)*

If the compound contains an acid hydrogen atom, the halogen and the remaining oxygen atoms are named as ligands, and the normal acid name format is used.

Example:

1. HSO_3Cl hydrogen chlorotrioxosulfate

If there are no acid hydrogen atoms, the compound is no longer an acid, and the name can be constructed by coordination nomenclature or by using the traditional name, if there is one, employing the root group (radical) of the parent acid, such as phosphoryl (PO), sulfuryl (SO_2), and thionyl (SO). See Section I-8.4.2.2 for a discussion of these group names. Of the two procedures, the former is preferred.

Examples:

2. SO_2Cl_2 dichlorodioxosulfur, or sulfuryl dichloride
3. $POCl_3$ trichlorooxophosphorus, or phosphoryl trichloride

When the compound to be named is based on a transition metal, and there are no acid hydrogen atoms, the compound should be named as a halide oxide (see Section I-8.5.4), and a coordination name is again preferred.

Note 9c. Names such as trithiocarbonic acid are not recommended.

Example:
4. $MoCl_2O_2$ dichlorodioxomolybdenum, or molybdenum dichloride dioxide

I-9.5.3.3   *Acid anhydrides*

Anhydrides of fully dehydrated inorganic acids are given oxide names, and anhydride names should no longer be used.

Example:
1. N_2O_5 dinitrogen pentaoxide (not nitric anhydride or nitric acid anhydride)

I-9.5.3.4   *Esters*

The naming of esters is not entirely consistent with the principles outlined above, because they are considered to be organic compounds and the recommendations of the *Nomenclature of Organic Chemistry*, 1979 edition, especially Section D, apply. The order of citation (organic substituent groups named first), the spacing between the substituent group name and the anion name, and the lack of a space between hydrogen and the anion name are especially to be noted. In every case there are coordination names, which are cited first below. The organic names are the last cited.

Examples:
1. $SO_2(OCH_3)_2$ dimethoxodioxosulfur, or dimethyl sulfate
2. $P(OCH_3)_3$ trimethoxophosphorus, or trimethyl trioxophosphate,
   or trimethyl phosphite
3. $HOSO_2OCH_3$ hydrogen methoxotrioxosulfate,
   or methyl hydrogentetraoxosulfate, or methyl hydrogensulfate

I-9.5.3.5   *Amides*

Names are generated using coordination principles. The anion NH_2^- as ligand is called 'amido'.

Example:
1. $HOSO_2NH_2$ hydrogen amidotrioxosulfate

This name is preferred to names such as sulfamidic acid (Note 9d), and abbreviated names such as sulfamide are also discouraged, although their importance in medicinal chemistry is well established. If the compound does not contain an acid hydrogen atom, the word 'acid' in the traditional name is replaced by amide. There is also a radicofunctional name analogous to that used for halides (Section I-9.5.3.2). Again, the coordination name is preferred. The radicofunctional name is cited last in Example 2 below.

Example:
2. $SO_2(NH_2)_2$ diamidodioxosulfur, or sulfuric diamide,
   or sulfuryl diamide

Note 9d. See the *Nomenclature of Organic Chemistry*, 1979 edition, Rule C-661, p. 243.

1-9.6   ACID NOMENCLATURE

This system is to be applied only to those oxoacids listed under 'acid nomenclature' in Table I-9.1. In this system of nomenclature, names consist of two words, the second of which is 'acid'. The first word has four constituents; the first constituent describes the ligands attached to the central atom, assuming that the acid hydrogen atom is fully ionized; the second is the stem of the central atom element name (Table I-9.2, column 2); the third is the ending -ic; the fourth, which is not always needed and may not be easy to define, is the oxidation number.

Examples:

1. H_2SO_4 tetraoxosulfuric acid
2. $HClO_3$ trioxochloric(v) acid
3. $HClO_4$ tetraoxochloric(VII) acid
4. $HMnO_4$ tetraoxomanganic(VII) acid
5. H_2MnO_4 tetraoxomanganic(VI) acid

This system does not indicate the number of acid hydrogen atoms, and should be restricted to the examples listed in Table I-9.1. Apart from the examples listed in Table I-9.1, which are allowed alternatives due to common usage, the names of oxoacids derived from transition metals should be named as hydroxooxometal compounds.

Example:

6. $[HReO_4]$ hydroxotrioxorhenium(VII)

The anion name derives straightforwardly from coordination nomenclature.

Example:

7. $[ReO_4]^-$ tetraoxorhenate(VII)

I-9.7   POLYNUCLEAR ACIDS

I-9.7.1   **Names of polynuclear acids without direct bonds between central atoms**

Polynuclear acids should be named using rules derived for coordination compounds. Exceptions to this, generally specified below, are allowed only to accord with established traditional practices.

The number of central atoms is indicated by using di-, tri-, etc. (Table III) to qualify the name stem listed in Table I-9.2. When the molecule contains more than one large unit of the same kind, the use of bis-, tris-, etc. (Table III) may permit the development of a simpler name. Due to the complexity of these acids, the number of acid hydrogen atoms should be specified. The anionic ligands are cited in alphabetical order. Any bridges (atom or group of atoms) between two central atoms are cited as ligands in alphabetical order, before the non-bridging anionic ligands, preceded by the descriptor μ and separated from the rest of the name by a hyphen. Two or more identical bridging groups are indicated by di-μ-, tri-μ-, etc. (see Chapters I-2 and I-10). Different bridging groups are listed in alphabetical order. When a bridge and an ordinary ligand have the same name, the bridging ligand is cited first.

Examples:
1. $H_4P_2O_7$ tetrahydrogen μ-oxo-hexaoxodiphosphate$(4-)$, or tetrahydrogen μ-oxo-bis(trioxophosphate)$(4-)$
2. $H_2S_2O_8$ dihydrogen μ-peroxo-hexaoxodisulfate$(2-)$, or dihydrogen μ-peroxo-bis(trioxosulfate)$(2-)$
3. $H_2S_4O_6$ dihydrogen μ-disulfido-bis(trioxosulfate)$(2-)$
4. $H_2[(O_2)_2OCrOOCrO(O_2)_2]$
                 dihydrogen μ-peroxo-bis(oxodiperoxochromate)$(2-)$

I-9.7.2   Polynuclear acids with direct bonds between central atoms

When two central atoms are directly linked, the presence of this particular linking bond may be shown at the end of the name and before the formal ionic charge by writing the italicized symbols of both central atoms separated by a long dash (compare coordination compounds containing a metal–metal bond) in parentheses.

Examples:
1. $H_2S_2O_6$ dihydrogen hexaoxodisulfate$(S\!-\!S)(2-)$
2. $(HO)_2OPPO(OH)_2$ tetrahydrogen hexaoxodiphosphate$(P\!-\!P)(4-)$

I-9.7.3   Isopolyacids (homopolyacids)

These materials are generally referred to in the literature as isopolyacids. The name homopolyacids is preferable because the Greek root of homo- implies 'the same', in direct contrast to that of hetero- signifying 'different', whereas the root of iso- implies equality. A detailed nomenclature of those compounds has been published elsewhere (Note 9e). The discussion presented here is brief and deals only with simple cases. Stoichiometric names can be simply constructed using hydrogen nomenclature (Section I-9.5).

Example:
1. $H_2Mo_6O_{19}$ dihydrogen nonadecaoxohexamolybdate$(2-)$

Acceptable abbreviated names may be given to polyoxoacids formally derived by condensation (with evolution of water) of units of the same mononuclear oxoacid, provided that the central atom of the mononuclear oxoacid has the highest oxidation state of the Periodic Group to which it belongs, that is, VI for sulfur, etc. The names are formed by indicating with numerical prefixes the number of atoms of central element present. It is not necessary to state the number of oxygen atoms.

Examples:
2. $H_2S_2O_7$ disulfuric acid, or dihydrogen disulfate
3. $H_2Mo_6O_{19}$ dihydrogen hexamolybdate
4. $H_6Mo_7O_{24}$ hexahydrogen heptamolybdate

Note 9e. See *Pure Appl. Chem.*, **59**, 1529 (1987).

Cyclic and chain structures may be distinguished by means of the italicized prefixes *cyclo-* and *catena-*, although the latter is usually omitted (as in Example 5).

Examples:

5. $H_5P_3O_{10}$ pentahydrogen triphosphate
6. $H_3P_3O_9$ trihydrogen *cyclo*-triphosphate
7. $H_3B_3O_6$ trihydrogen *cyclo*-tri-μ-oxo-tris(oxoborate)

To name simple, linear isopolyacids (homopolyacids) in which an oxygen atom is replaced by another ligand, a numbering procedure is needed. The central atoms are numbered from one end of the chain so that the atom(s) bearing substituents carry the lowest possible locants. This numbering principle is well established in organic nomenclature.

Examples:

8. $H_5P_3O_9S$ pentahydrogen di-μ-oxo-heptaoxo-1-thiotriphosphate(5−)
9. $H_5P_3O_8S_2$ pentahydrogen di-μ-oxo-hexaoxo-1,2-dithiotriphosphate(5−)
10. $H_4P_3O_9NH_2$ tetrahydrogen 1-amido-di-μ-oxo-heptaoxotriphosphate(4−)
11. $H_6P_4O_{12}(NH)$ hexahydrogen 1,2-μ-imido-di-μ-oxo-decaoxotetraphosphate(6−)

More complicated structures will be discussed in a document in preparation (1989) on chain and ring compounds. See also Note 9e.

I-9.7.4   **Heteropolyacids**

A detailed nomenclature of these compounds, has been presented elsewhere (Note 9e). The discussion here treats only simple cases. Names are developed using coordination nomenclature (Chapter I-10). Thus, the central element names are placed at the end of the name of the acid, in alphabetical order, without spaces, and in parentheses; then the ending -ate is added.

Example:

1. H_3PSO_7 trihydrogen μ-oxo-hexaoxo(phosphorussulfur)ate(3−)

When one central atom is a transition metal atom and the other not, then the latter is treated as a ligand to the former.

Example:

2. $H_2CrO_3(SO_4)$ dihydrogen trioxo(tetraoxosulfato)chromate(2−)

In more complicated heteropolyacids with many central atoms, the heteroatoms are surrounded either tetrahedrally or octahedrally by oxygen atoms. Where the structure is not known and the heteroatom is a Main Group element, the heteroatom and its shell of oxygen atoms are named as a ligand, just as in the dinuclear case cited above, Example 2.

Examples:

3. $H_4SiW_{12}O_{40}$
   tetrahydrogen hexatriacontaoxo(tetraoxosilicato)dodecatungstate$(4-)$
4. $H_6P_2W_{18}O_{62}$
   hexahydrogen tetrapentacontaoxobis(tetraoxophosphato)octadecatungstate$(6-)$

Some abbreviated semi-trivial names are retained for present use due to long-standing usage. This applies if all the central atoms are the same, if the polyanion contains only oxygen atoms as ligands and only one kind of heteroatom, and if the oxidation state of the central atoms corresponds to the highest oxidation state of the Periodic Group in which they occur. In this usage, the Main Group atoms receive specific abbreviated names for incorporation into the hetero-polyacid name. These are shown immediately below.

B boro	Si silico	Ge germano
P phospho	As arseno	

Examples:

5. $H_4SiW_{12}O_{40}$ tetrahydrogen silicododecatungstate
6. $H_6P_2W_{18}O_{62}$ hexahydrogen diphosphooctadecatungstate

I-9.8   IONS DERIVED FROM OXOACIDS

I-9.8.1   Anions

The hydrogen nomenclature name described above (Section I-9.5) consists of two parts, the second of which is an anion name (Table I-9.2). This can stand alone to represent the anion itself.

Examples:

1. H_2SO_4 dihydrogen tetraoxosulfate$(2-)$
   SO_4^{2-} tetraoxosulfate$(2-)$
2. $H_2[OsO_4]$ dihydrogen tetraoxoosmate$(2-)$
   $[OsO_4]^{2-}$ tetraoxoosmate$(2-)$

If all the acid hydrogen atoms are not removed (as in acid salts), the word hydrogen together with any necessary multiplicative prefix is joined to the second part of the 'acid' name and the charge number is changed appropriately.

Example:

3. HSO_4^- hydrogentetraoxosulfate$(1-)$, or
                      monohydrogentetraoxosulfate

Traditional names are still accepted for the exceptions listed in Table I-9.1. The ending -ic of the acid name becomes -ate in the anion name, and -ous becomes -ite.

Examples:

4.	H_2SO_4	sulfuric acid
	SO_4^{2-}	sulfate
5.	H_2SO_3	sulfurous acid
	SO_3^{2-}	sulfite
6.	$H_2S_2O_6$	dithionic acid
	$S_2O_6^{2-}$	dithionate
7.	HIO_3	iodic acid
	IO_3^{-}	iodate

Further discussion can be found in Section I-8.3.3.8.

I-9.8.2   **Cations**

The cations considered here are obtained by adding formally one or more hydrogen cations (hydrons) to a neutral molecule of the acid (see Section I-8.2.3.5).

Examples:

1.	$(H_3SO_4)^+$	trihydroxooxosulfur(VI) cation
2.	$(H_2NO_3)^+$	dihydroxooxonitrogen(V) cation

Note that an extension of the organic style of nomenclature (cf. the *Nomenclature of Organic Chemistry*, 1979 edition, Rule C-82, p. 134) as in $(CH_3CO_2H_2)^+$ = ethanoic acidium, is discouraged because it is based on the word 'acid' and is often not easily adaptable to languages other than English.

I-9.9   SPECIAL CASE OF MONONUCLEAR PHOSPHORUS AND ARSENIC OXOACIDS

I-9.9.1   **General remarks**

Phosphorus oxoacids have many derivatives that are often considered to be organic compounds. Several systems of nomenclature have been developed over the years and they are still more-or-less widely used. Although these nomenclatures are described in great detail in the *Nomenclature of Organic Chemistry*, 1979 edition, pp. 382–405, this Chapter dealing with nomenclature of inorganic oxoacids would not be complete without describing the basic principles. The various systems in use are examples of substitutive nomenclature, infix and prefix replacement nomenclature, and coordination nomenclature. These are described briefly below.

I-9.9.2   **Substitutive nomenclature**

Substitutive nomenclature is the principal system used in organic chemistry, and it is treated *in extenso* in the *Nomenclature of Organic Chemistry*, 1979 edition, Sections C-0.1 and D-1.3. The fundamental device involves regarding compounds as obtained by substitution of hydrogen atoms of parent compounds, generally hydrides, by other

groups, the substituents. In organic chemistry, the parent hydrides are usually hydrocarbons or heterocyclic compounds. In the present context, the parents are hydrides of
phosphorus or arsenic. The mononuclear parent acids are phosphonic acid, $HPO(OH)_2$,
phosphinic acid, $H_2PO(OH)$, arsonic acid, $HAsO(OH)_2$, and arsinic acid, $H_2AsO(OH)$.
Similar names are used for the arsenic(III) and phosphorus(III) acids (Section I-9.10.1).
Note that electronegative groups such as Cl are regarded as replacing an OH rather than
an H, so that $HPO(Cl)(OH)$ is a derivative of phosphonic acid, not of phosphinic acid.
For a fuller discussion see the *Nomenclature of Organic Chemistry*, 1979 edition,
Sections D-5.5, p. 395 and D-5.6, p. 399.

Examples:
1. $(C_6H_5)_2PO(OH)$ diphenylphosphinic acid
2. $(H_2NC_6H_4)AsO(OH)_2$ (aminophenyl)arsonic acid

I-9.9.3   Infix and prefix replacement nomenclature

This system of nomenclature can be used for mononuclear phosphorus and arsenic
oxoacids and their anions in which replacement of oxygen or an hydroxy group has
occurred. Some of the infixes used are listed below (see also the *Nomenclature of Organic
Chemistry*, 1979 edition, p. 382).

Infix	Replacing atom or group	Replaced atom or group
-peroxo-	$-OO-$	$-O-$
-thio-	$=S, -S-$	$=O, -O-$
-seleno-	$=Se, -Se-$	$=O, -O-$
-telluro-	$=Te, -Te-$	$=O, -O-$
-amid(o)-	$-NH_2$	$-OH$
-imid(o)-	$=NH$	$=O$
-nitrid(o)-	$\equiv N$	$=O$ and $-OH$
-fluorid(o)-	$-F$	$-OH$
-chlorid(o)-	$-Cl$	$-OH$
-bromid(o)-	$-Br$	$-OH$
-azid(o)-	$-N_3$	$-OH$
-iodid(o)-	$-I$	$-OH$
-cyanid(o)-	$-CN$	$-OH$

The appropriate infix is inserted before the suffix -ic, -oic, or -ous of the parent acid
name, or -ate, -oate, or -ite of the parent anion name and the name of the parent acid is
modified as below. The 'o' in parentheses in the above list is omitted for grammatical
reasons in certain cases. If more than one infix is necessary, they are cited in alphabetical
order, any multiplicative prefixes required being neglected for this ordering purpose.

phosphinic acid	$H_2PO(OH)$	phosphin(o)-
phosphonic acid	$HPO(OH)_2$	phosphon(o)-
phosphoric acid	H_3PO_4	phosphor(o)-

Similar modifications apply to the names of the arsenic acids.

Example:

1. $(C_2H_5)_2PS(SH)$ diethylphosphinodithioic acid

Replacement of an oxygen or hydroxyl in an oxoacid of phosphorus(v) or arsenic(v) can also be indicated using prefixes.

Prefixes	Replacing atom or group	Replaced atom or group
thio-	$=S$	$=O$
amido-	$-NH_2$	$-OH$
fluoro-	$-F$	$-OH$
chloro-	$-Cl$	$-OH$
bromo-	$-Br$	$-OH$
iodo-	$-I$	$-OH$
imido-	$=NH$	$=O$

Where more than one prefix is cited, alphabetical ordering as for infixes is used.

Examples:

2. H_3PO_3S thiophosphoric acid
3. $(C_6H_5)_2P(NH)(OH)$ diphenylimidophosphinic acid

An alternative approach to the nomenclature of these compounds would be to use the λ-convention, for which see *Pure Appl. Chem.*, **56**, 769 (1984).

I-9.9.4   **Coordination nomenclature**

These compounds may all be named by coordination nomenclature with phosphorus and arsenic (and also antimony and bismuth) considered as central atoms.

Examples:

1. $H[P(C_2H_5)_2S_2]$ hydrogen diethyldithiophosphate(v)
2. $[PO(OCH_3)_3]$ trimethoxooxophosphorus

I-9.10   NAMES FOR OXOACIDS CONTAINING PHOSPHORUS OR ARSENIC, AND THEIR DERIVATIVES

I-9.10.1   **Compounds with coordination number 3 which may be considered as oxoacids or derivatives**

Compounds with coordination number 3 can be named by three methods. Coordination nomenclature can be used based on phosphorus, arsenic, antimony, or bismuth as central atom (a). Alternatively, they can be treated as derivatives of the parent acids listed below (b).

$H_2P(OH)$ phosphinous acid
$HP(OH)_2$ phosphonous acid

$P(OH)_3$	phosphorous acid
$H_2As(OH)$	arsinous acid
$HAs(OH)_2$	arsonous acid
$As(OH)_3$	arsenous acid

Finally, they can be considered as substitution products of MH_3 (M = P, As, Sb, or Bi) (c).

Examples 1–3 below demonstrate the three kinds of name so developed.

Examples:

1. $(C_6H_5)_2P(OCH_3)$ a methoxodiphenylphosphorus(III)
                           b methyl diphenylphosphinite
                           c methoxydiphenylphosphane (Note 9f)

2. $(C_6H_5)P(OH)_2$ a dihydroxophenylphosphorus(III)
                           b phenylphosphonous acid
                           c dihydroxy(phenyl)phosphane (Note 9f)

3. $P(OCH_3)_3$ a trimethoxophosphorus(III)
                           b trimethyl phosphite
                           c trimethoxyphosphane (Note 9f)

When more than one acid group, each containing one atom of trivalent phosphorus or arsenic is present as principal organic group in an organic compound, the compound may be named as a derivative of a parent acid substituted by a polyvalent group (b). Variants (a) and (c) in Example 4 use substitutive and coordination nomenclatures as described above. Derivatives of these acids are named by applying the same principles.

Example:

4. 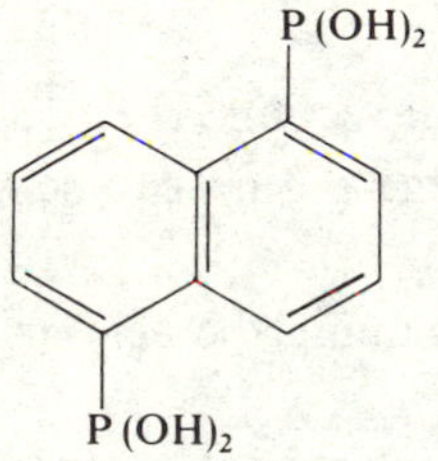   a μ-1,5-naphthylenebis[dihydroxophosphorus(III)]
     b 1,5-naphthylenebis(phosphonous acid)
     c P,P,P',P'-tetrahydroxy-1,5-naphthylenebis(phosphane) (Note 9f)

I-9.10.2   Oxoacids of pentavalent phosphorus (or arsenic) containing carbon directly linked to phosphorus

These compounds are named as substitution derivatives of the following parent acids.

$HPO(OH)_2$	phosphonic acid
$H_2PO(OH)$	phosphinic acid
$HAsO(OH)_2$	arsonic acid
$H_2AsO(OH)$	arsinic acid

Note 9f. Phosphane is the recommended name for PH_3. Phosphine is retained only for a few special cases (Chapter I-7).

Examples:

1. $(C_6H_5)_2PO(OH)$ diphenylphosphinic acid

$PO(OH)_2$

2. 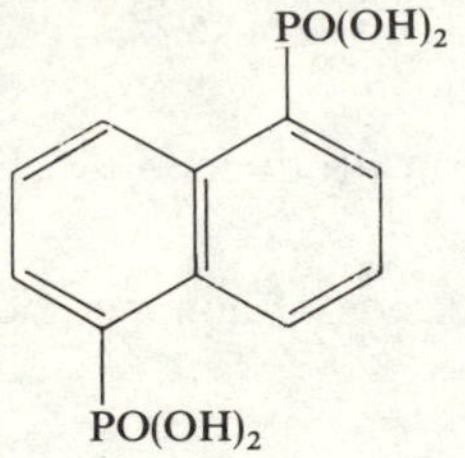 1,5-naphthylenebis(phosphonic acid)

$PO(OH)_2$

Organic nomenclature has priority rules (the *Nomenclature of Organic Chemistry*, 1979 edition, Section C-0.2, p. 112) which may necessitate that the oxoacid be treated as a substituent in an organic molecule. It would then be designated by one of the following prefixes. The names for the arsenic analogues are similar.

$-PO(OH)_2$	phosphono
$=PO(OH)$	phosphinico
$-PO_3{}^{2-}$	phosphonato
$=PO_2{}^-$	phosphinato

Examples:

3. $(HO)_2OPCH_2COOH$ phosphonoacetic acid

4. 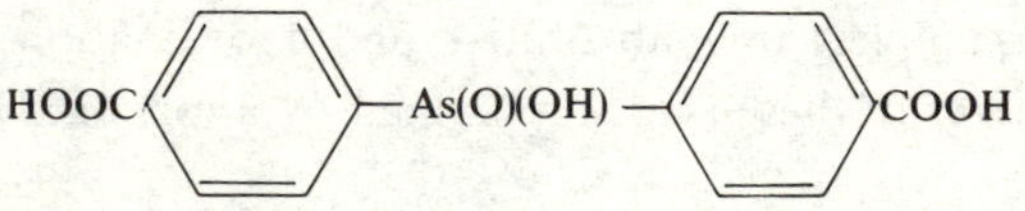 $HOOC$—⟨⟩—$As(O)(OH)$—⟨⟩$COOH$ 4,4'-arsinicobis(benzoic acid)

5. $[O_3P(C_6H_4COO)]^{3-}$ 2-phosphonatobenzoate

Derivatives of the acids listed at the beginning of Section I-9.10.2 and in which oxygen atoms and/or hydroxyl groups are replaced by other groups can be named in one of the following ways.

(a) By use of coordination nomenclature or by naming the acid as a hydrogen salt in hydrogen nomenclature (Table I-9.1).

(b) By consideration of them as replacement derivatives of the acids named according to hydrogen nomenclature (Section I-9.5).

(c) By use of infix or prefix replacement nomenclature (see Section I-9.9.3).

Whichever method is used, groups of equivalent status are always listed alphabetically by initial letters, omitting consideration of any multiplicative prefixes for this purpose. In the Examples 6 and 7, below, a, b and c refer to the nomenclature methods specified immediately above this paragraph.

Examples:

6. $(C_2H_5)_2P(S)(SH)$ a diethyl(hydrogensulfido)thiophosphorus(v)

   b hydrogen diethyldithiophosphate(v)

   c diethyldithiophosphinic acid

   c diethylphosphinodithioic acid

7. $C_6H_5P(O)Cl(OH)$ a chlorohydroxooxophenyl phosphorus(v)

                                            b hydrogen chlorodioxophenylphosphate(v)

                                            c phenyl(chloro)phosphonic acid

                                            c phenylphosphonochloridic acid

The point of attachment of the acid hydrogen atom is not generally specified in naming acids because they often exist as a mixture of tautomeric forms in solution; in the solid state the hydrogen atom is usually bonded to more than one anion. However, it may be desired to designate a particular tautomer; this can be done by designating the atom(s) to which the acid hydrogen atom(s) are bound (b and c in Examples 8 and 9). When using coordination nomenclature, the hydrogen atom is considered part of a ligand, for instance, OH and SH, and the compound is named accordingly (a in Examples 8 and 9).

Examples:

8. $C_2H_5P(Se)(OH)_2$ a ethyldihydroxoselenidophosphorus(v)

                                            b ethylselenophosphonic *O,O'*-acid

                                            c ethylphosphonoselenoic *O,O'*-acid

9. $C_2H_5P(O)(OH)(SeH)$ a ethyl(hydrogenselenido)hydroxooxophosphorus(v)

                                            b ethylselenophosphonic *O,Se*-acid

                                            c ethylphosphonoselenoic *O,Se*-acid

Anions may be named by dropping the word acid from the acid name and changing the -ic ending to -ate. If necessary the formal ionic charge may be added.

I-9.10.3   **Derivatives of oxoacids containing pentavalent phosphorus or arsenic**

In each case below the general naming procedure is indicated. The letters a, b, and c designate the nomenclature methods as presented in Section I-9.10.2, p. 140.

    Esters: the ending -ic acid of the acid name is changed to -ate and the name qualified by the names of the organic groups involved. Phosphoric acid is a minor exception, since it gives rise to phosphate rather than phosphorate.

Examples:

1. $PO(OCH_3)_3$ a trimethoxooxophosphorus(v)

                                            b trimethyl phosphate

2. $C_6H_5PH(S)(OCH_3)$ a hydridomethoxo(phenyl)thiophosphorus(v)

                                            b *O*-methyl phenylphosphinothioate

                                            c *O*-methyl phenylthiophosphinate

3. $C_2H_5P(O)(OC_2H_5)OH$ a ethoxoethylhydroxooxophosphorus(v)

                                          b,c ethyl hydrogen ethylphosphonate

    Amides: the word acid in the acid name is changed to amide.

Example:

4. $(CH_3)_2P(O)[NH(CH_3)]$ a (methylamido)dimethyloxophosphorus

                                              c *N,P,P*-trimethylphosphinic amide

Where there is more than one amido-group, the appropriate multiplicative prefix qualifies its name.

Acid halides (halogenides): the word acid in the acid name is replaced by the appropriate halide name.

Example:

5. $(C_2H_5)_2PCl(NC_6H_5)$ a chloro(diethyl)(phenylimido)phosphorus

                                       c *P*,*P*-diethyl-*N*-phenylphosphinimidic chloride,

                                              or diethyl(phenylimido)phosphinic chloride

Mixed anhydrides: the *Nomenclature of Organic Chemistry*, 1979 edition, rule C-491.3 suggests ordering the first parts of the acid names alphabetically and following them by the word anhydride. Example 6 shows this, and also a coordination name.

Example:

6. $CH_3C(O)-OPO(OH)_2$ acetatodihydroxooxophosphorus (coordination name),

                                        or acetic phosphoric monoanhydride (organic name)

In substitutive nomenclature, the priority rules may require that the acid be treated as a substituent in an organic parent. The group names required for these purposes are shown below. Similar names apply to arsenic derivatives. Where oxygen is replaced by nitrogen or sulfur, then the groups are best named by prefix or infix replacement nomenclature.

$H_2P(O)-$	phosphinoyl
$H(O)P<$	phosphonoyl
$(O)P\lessdot$	phosphoryl
$H_2P(S)-$	thiophosphinoyl, or phosphinothioyl
$HP(N)-$	nitridophosphonoyl, or phosphononitridoyl

An instance of the application of substitutive nomenclature is shown in Example 7,a. There is also the alternative of using an organometallic radical or substituent name (see Chapter I-8 and Table VII). This is shown in Example 7,b.

Example:

7. $(C_2H_5O)_2P(O)C_6H_4COOH$ a 4-(diethoxyphosphoryl)benzoic acid

                                             b 4-diethoxyoxophosphoriobenzoic acid

I-9.11   FINAL REMARKS

This Chapter has discussed the various kinds of nomenclature which are available for naming oxoacids. It presents more than one name for many compounds, and consequently often allows the user the choice. However, the Commission wishes to encourage coordination-based inorganic nomenclature, and to discourage trivial systems which need to be learned by rote.

The principles enunciated may not be easy to use with the more complex systems and with the heteropolyacid derivatives which are increasingly the subject of research. A document on rings and chains is currently (1989) being prepared by the Commission and one on polyanions has already appeared [*Pure Appl. Chem.*, **59**, 1529 (1987)]. Specialized documents such as these should be used in conjunction with this Chapter.

I-10 Coordination Compounds

CONTENTS

I-10.1 INTRODUCTION

The general and fundamental definitions necessary for formulating and naming coordination compounds are presented in this Chapter, including reference to their origins and foundations. Coordination entity, central atom, ligand, coordination polyhedron, coordination number, chelation, and bridging ligands are first defined and the roles of oxidation number and additive nomenclature are explained. These are used to develop rules for the formulae and names of mononuclear coordination compounds with monodentate ligands. Stereochemical descriptors are introduced to distinguish diastereoisomeric structures for many coordination polyhedra. Rules for naming derivatives of polydentate ligands and the presentation of chirality descriptors then follow. Finally, organometallic compounds, bridged structures, and metal clusters are then treated in preliminary fashion.

I-10.2 CONCEPTS AND DEFINITIONS

The defining concepts for coordination compounds rest on the historically significant concepts of primary and secondary valence. Primary valences were obvious from the stoichiometries of simple compounds such as SO_2, $NiCl_2$, $Fe_2(SO_4)_3$, and $PtCl_2$. Recognition that the addition to these of other, independently stable substances, such as H_2O, NH_3, and KCl, produced new materials led to the formulation and naming of coordination compounds. Such behaviour was frequently observed among compounds of the metallic elements, e.g., $NiCl_2 \cdot 4H_2O$, $Co_2(SO_4)_3 \cdot 12NH_3$, or $PtCl_2 \cdot 2KCl$. Such species were considered characteristic of certain metallic elements and were called complex compounds, in recognition of the stoichiometric complications that they represented. Each coordination compound either is, or contains, a coordination entity (or complex) the composition, structure, and chemistry of which reflect what was eventually defined as the secondary valence of the metallic element. While the concepts in the preceding discussion have been applied to compounds of the metals, it is recognized that they are often also useful in describing other compounds.

I-10.2.1 Coordination entity

A *coordination entity* is composed of a central atom, usually that of a metal, to which is attached a surrounding array of other atoms or groups of atoms, each of which is called a ligand. Classically, a ligand was said to satisfy either a secondary or a primary valence of the central atom and the sum of these valences (often equal to the number of ligands) was called the coordination number (see below). In formulae, the coordination entity is enclosed in square brackets whether it is charged or uncharged.

Examples:
1. $[Co(NH_3)_6]^{3+}$
2. $[PtCl_4]^{2-}$
3. $[Fe_3(CO)_{12}]$

I-10.2.2   **Central atom**

The *central atom* is the atom in a coordination entity which binds other atoms or groups of atoms (ligands) to itself, thereby occupying a central position in the coordination entity. The central atoms in $[NiCl_2(H_2O)_4]$, $[Co(NH_3)_6]^{3+}$, and $[PtCl_4]^{2-}$ are nickel, cobalt, and platinum, respectively.

I-10.2.3   **Ligands**

The *ligands* are the atoms or groups of atoms bound to the central atom. The root of the word is often converted into other forms, such as *to ligate*, meaning to coordinate as a ligand, and the derived participles, *ligating* and *ligated*.

I-10.2.4   **Coordination polyhedron**

It is standard practice to think of the ligand atoms that are directly attached to the central atom as defining a *coordination polyhedron* (or polygon) about the central atom. Thus $[Co(NH_3)_6]^{3+}$ is an octahedral ion and $[PtCl_4]^{2-}$ is a square planar ion. In this way the coordination number may equal the number of vertices in the coordination polyhedron. Exceptions are common among organometallic compounds (see Section I-10.9).

Examples:

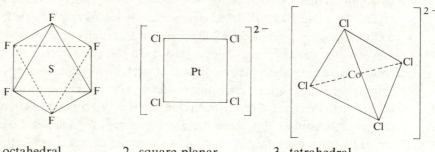

1. octahedral   2. square planar   3. tetrahedral
   coordination   coordination   coordination
   polyhedron   polygon   polyhedron

Historically, the concepts and nomenclature of coordination compounds were unambiguous for a long time, but complications have arisen more recently. According to tradition, every ligating atom or group was recognized as bringing one lone-pair of electrons to the central atom in the coordination entity. This sharing of ligand electron pairs became synonymous with the verb 'to coordinate.' Further, in the inevitable electron bookkeeping that ensues upon consideration of a chemical compound, the coordination entity was dissected (in thought) by removal of each ligand in such a way that each ligating atom or group took two electrons with it. Coordination number is simply definable when such a thought process is applied.

I-10.2.5   **Coordination number**

As defined for typical coordination compounds, the coordination number equals the number of sigma-bonds between ligands and the central atom. Even though simple ligands such as CN^-, CO, N_2, and $P(CH_3)_3$ may involve both sigma- and pi-bonding between the ligating atom and the central atom, the pi-bonds are not considered in determining the coordination number (Note 10a).

The sigma-bonding electron pairs in the following examples are indicated by : before the ligand formulae.

Examples:

	Complex	*Coordination number*
1.	$[Co(:NH_3)_6]^{3+}$	6
2.	$[Fe(:CN)_6]^{3-}$	6
3.	$[Ru(:NH_3)_5(:N_2)]^{2+}$	6
4.	$[Ni(:CO)_4]$	4
5.	$[Cr(:CO)_5]^{2-}$	5
6.	$[Co(:Cl)_4]^{2-}$	4

I-10.2.6   **Chelation**

Chelation involves coordination of more than one sigma-electron pair donor group from the same ligand to the same central atom. The number of such ligating groups in a single chelating ligand is indicated by the adjectives didentate (Note 10b), tridentate, tetradentate, pentadentate, etc. (see Table III for a list of numerical prefixes). The number of donor groups from a given ligand attached to the same central atom is called the denticity.

Examples:

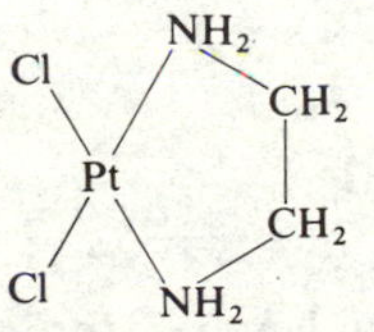

1. didentate chelation

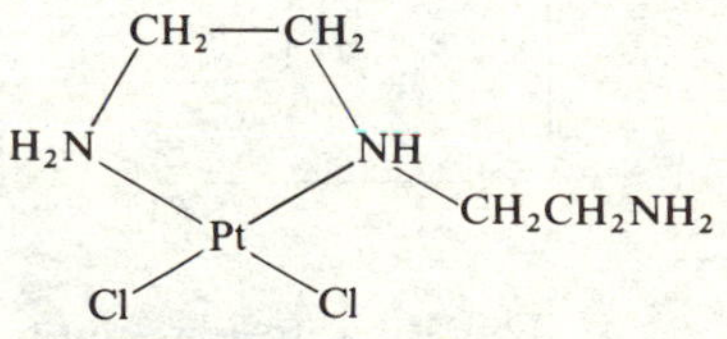

2. didentate chelation

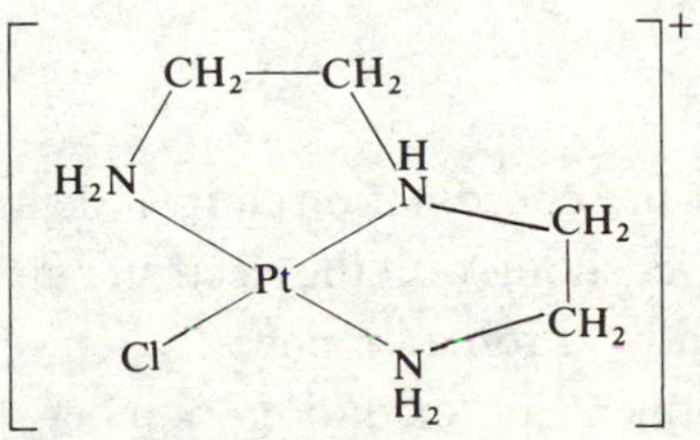

3. tridentate chelation

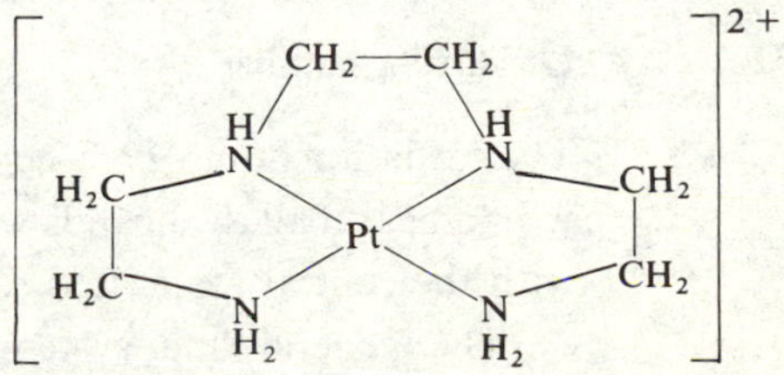

4. tetradentate chelation

Note 10a. This definition is appropriate to coordination compounds, but not necessarily to other areas, such as crystallography.

Note 10b. Previous versions of the *Nomenclature of Inorganic Chemistry* used bidentate rather than didentate, for linguistic consistency. In this edition, a single set of prefixes is used throughout and these are listed in Table III.

Almost from the beginning of coordination chemistry, it had been realized that many polyfunctional molecules and ions (e.g., 1,2-ethanediamine, N-(2-aminoethyl)-1,2-ethanediamine, N,N'-bis(2-aminoethyl)-1,2-ethanediamine, oxalate, and glycinate) are capable of simultaneous use of more than one atom to bind to central atoms. Free 1,2-ethanediamine is a difunctional molecule while, for example, N,N'-bis(2-aminoethyl)-1,2-ethanediamine is tetrafunctional. Each can use all of its functional groups simultaneously to coordinate to a single metal ion. When this happens, cyclic structures called chelate rings are formed and this process of simultaneous coordination is called chelation.

The number of points of coordination to a single metal ion by a polyfunctional molecule is denoted by an adjective derived from the root -dentate. 1,2-Ethanediamine can chelate in a didentate manner while N,N'-bis(2-aminoethyl)-1,2-ethanediamine can act as a tetradentate chelating agent. The prefixes given in Table III are used: di-, tri-, tetra-, penta-, hexa-, hepta-, octa-, nona-, deca-, undeca-, etc., and finally, poly-.

If a 1,2-ethanediamine molecule coordinates to two metal ions, it does NOT chelate but *coordinates* in a *monodentate* fashion to each metal ion, forming a connecting link or bridge. Olefinic, aromatic, and other unsaturated molecules attach to central atoms using some or all of their multiply bonded atoms. Complexes of this kind deviate in clearly definable ways from the classic patterns of coordination chemistry (see Section I-10.9).

Examples:

1. $[(NH_3)_5Co-NH_2CH_2CH_2NH_2-Co(NH_3)_5]^{6+}$
   bis-monodentate *coordination* (not chelation) by 1,2-ethanediamine

2.
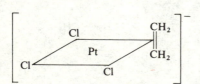

This dihapto bonding (see Section I-10.9) is monodentate coordination.

I-10.2.7   **Oxidation number**

The oxidation number of a central atom in a coordination entity is defined as the charge it would bear if all the ligands were removed along with the electron pairs that were shared with the central atom. It is represented by a roman numeral.

The general and systematic treatment of oxidation number follows from the application of the classical definition of coordination number. It must be emphasized that oxidation number is an index derived from a simple and formal set of rules (Section I-5.5.2.2) and that it is not a direct indicator of electron distribution. In certain cases, the formalism does not give acceptable central atom oxidation numbers. In such ambiguous cases, the net charge on the coordination entity is preferred in most nomenclature practices. In the Examples that follow, the relationship of oxidation number to coordination number is illustrated.

Examples:

Complex	Ligand list	Central atom oxidation number
1. $[Co(NH_3)_6]^{3+}$	$6 NH_3$	III
2. $[CoCl_4]^{2-}$	$4 Cl^-$	II
3. $[MnO_4]^-$	$4 O^{2-}$	VII
4. $[MnFO_3]$	$3 O^{2-} + 1 F^-$	VII
5. $[Co(CN)_5H]^{3-}$	$5 CN^- + 1 H^-$	III
6. $[Fe(CO)_4]^{2-}$	$4 CO$	$-II$
7. $[PtCl_2(C_2H_4)(NH_3)]$	$2 Cl^- + NH_3 + C_2H_4$	II

I-10.2.8 **Coordination nomenclature, an additive nomenclature**

According to a useful, historically-based formalism, coordination compounds are considered to be produced by addition reactions and so they were named on the basis of an additive principle. The name is built up around the central atom name, just as the coordination entity is built up around the central atom.

Example:
1. Addition of ligands to a central atom:
   $$Ni^{2+} + 6H_2O \rightarrow [Ni(H_2O)_6]^{2+}$$
   Addition of ligand names to a central atom name:
   hexaaquanickel(II) ion

This nomenclature extends to still more complicated structures where central atoms are added together to form dinuclear, trinuclear, and even polynuclear species from some mononuclear building blocks. The persistent centrality of the central atom is emphasized by the root -nuclear.

I-10.2.9 **Bridging ligands**

In polynuclear species it is necessary to distinguish yet another ligand behaviour, the action of the ligand as a bridging group.

A *bridging ligand* bonds to two or more central atoms simultaneously. Thus, bridging ligands link central atoms together to produce coordination entities having more than one central atom. The number of central atoms joined into a single coordination entity by bridging ligands or metal–metal bonds is indicated by dinuclear, trinuclear, tetranuclear, etc.

The *bridge index* is the number of central atoms linked by a particular bridging ligand (Section I-10.8.2). Bridging can be through one atom or through a longer array of atoms.

Examples:
1. $[(NH_3)_5Co–Cl–Ag]^{3+}$

2.

3.

I-10.2.10   **Metal–metal bonds**

Isolated metal–metal bonds in simple structures (Section I-10.8.3.1) are readily incorporated into coordination nomenclature, but complications arise when structures which involve three or more central atoms are considered. The cluster of central atoms in such species is treated in Section I-10.8.3.3.

Examples:
1. $[Br_4Re–ReBr_4]^{2+}$
2. $[(CO)_5Re–Co(CO)_4]$

I-10.3   FORMULAE FOR MONONUCLEAR COORDINATION
             COMPOUNDS WITH MONODENTATE LIGANDS

I-10.3.1   **Sequence of symbols within the coordination formula**

The central atom is listed first. The formally anionic ligands appear next and they are listed in alphabetic order according to the first symbols of their formulae, as required in Section I-4.6.7. The neutral ligands follow, also in alphabetical order, according to the same principle. Polydentate ligands are included in the alphabetical list, according to Section I-4.6.1.3.

Different applications often require considerable flexibility in the writing of formulae. For example, while Section I-4.6.7 implies a certain way of writing ligand formulae, they need to be so written only during the determination of the position of the ligand in the coordination formula. Considerable precedent exists for displaying the ligand formula with the donor atom nearest the central atom. Ligand structural formulae are sometimes needed. Complicated organic ligands may be designated in formulae with abbreviations (see Section I-10.4.5.7). The Examples in Section I-10.3.2 illustrate good common usage.

I-10.3.2   **Uses of enclosing marks**

The formula for the entire coordination entity, whether charged or not, is enclosed in square brackets. When ligands are polyatomic, their formulae are enclosed in parentheses. Ligand abbreviations are also enclosed in parentheses. In the special case of coordination entities, the nesting order of enclosures is as given in Sections I-2.2 and I-4.6.7. There should be no space between representations of ionic species within a coordination formula.

Examples:
1. $[Co(NH_3)_6]Cl_3$
2. $[CoCl(NH_3)_5]Cl_2$
3. $[CoCl(NO_2)(NH_3)_4]Cl$
4. $[PtCl(NH_2CH_3)(NH_3)_2]Cl$
5. $[CuCl_2\{O=C(NH_2)_2\}_2]$
6. $K_2[PdCl_4]$
7. $K_2[OsCl_5N]$
8. $Na[PtBrCl(NO_2)(NH_3)]$
9. $[Co(en)_3]Cl_3$

I-10.3.3   Ionic charges and oxidation numbers

If the formula of a charged coordination entity is to be written without that of the counter-ion, the charge is indicated outside the square bracket as a right superscript, with the number before the sign. The oxidation number of a central atom may be represented by a roman numeral used as a right superscript on the element symbol.

Examples:
1. $[PtCl_6]^{2-}$
2. $[Cr(H_2O)_6]^{3+}$
3. $[Cr^{III}(NCS)_4(NH_3)_2]^{-}$
4. $[Cr^{III}Cl_3(H_2O)_3]$
5. $[Fe^{-II}(CO)_4]^{2-}$

I-10.4   NAMES FOR MONONUCLEAR COORDINATION COMPOUNDS WITH MONODENTATE LIGANDS

I-10.4.1   Sequence of central atom and ligand names

The ligands are listed in alphabetical order, without regard to charge, before the names of the central atom. Numerical prefixes indicating the number of ligands are not considered in determining that order (see Section I-5.5.1).

Example:
1. di*chl*oro(*di*phenylphosphine)(*t*hiourea)platinum(II).

I-10.4.2   Number of ligands in a coordination entity

Two kinds of numerical prefix are available for indicating the number of each kind of ligand within the name of the coordination entity (see Table III). The simple di-, tri-, etc., derived from cardinal numerals, are generally recommended. The prefixes bis-, tris-, tetrakis-, etc., derived from ordinals, are used with complex expressions and when required to avoid ambiguity; for example, one would use diammine but bis(methylamine) to make a distinction from dimethylamine. When the latter multiplicative prefixes are used, enclosing marks are placed around the multiplicand. Enclosing marks are not required with the simpler prefixes di-, tri-, etc. For the nesting order of enclosing marks, see Chapter I-2. There is no elision of vowels or use of a hyphen in tetraammine and similar names, except for compelling linguistic reasons.

I-10.4.3   Terminations for names of coordination entities

All anionic coordination entities take the ending -ate, whereas no distinguishing termination is used for cationic or neutral coordination entities.

I-10.4.4   **Charge numbers, oxidation numbers, and ionic proportions**

When the oxidation number of the central atom can be defined without ambiguity, it may be indicated by appending a roman numeral to the central atom name (Note 10c). This number is enclosed in parentheses after the part of the name denoting the central atom. No positive sign is used. When necessary a negative sign is placed before the number. Arabic zero indicates the zero oxidation number. No space is left between this number and the rest of the name.

Alternatively, the charge on a coordination entity may be indicated. The net charge is written in arabic numbers on the line, with the number preceding the charge sign, and enclosed in parentheses. It follows the name of the central atom without the intervention of a space (Note 10d).

The stoichiometric proportions of ionic entities may be given by using stoichiometric prefixes on both ions, as necessary.

Examples:
1. $K_4[Fe(CN)_6]$
   potassium hexacyanoferrate(II)
   potassium hexacyanoferrate(4−)
   tetrapotassium hexacyanoferrate
2. $[Co(NH_3)_6]Cl_3$
   hexaamminecobalt(III) chloride
3. $[CoCl(NH_3)_5]Cl_2$
   pentaamminechlorocobalt(2+) chloride
4. $[CoCl(NO_2)(NH_3)_4]Cl$
   tetraamminechloronitrito-*N*-cobalt(III) chloride
5. $[PtCl(NH_2CH_3)(NH_3)_2]Cl$
   diamminechloro(methylamine)platinum(II) chloride
6. $[CuCl_2\{O{=}C(NH_2)_2\}_2]$
   dichlorobis(urea)copper(II)
7. $K_2[PdCl_4]$
   potassium tetrachloropalladate(II)
8. $K_2[OsCl_5N]$
   potassium pentachloronitridoosmate(2−)
9. $Na[PtBrCl(NO_2)(NH_3)]$
   sodium amminebromochloronitrito-*N*-platinate(1−)
10. $[Fe(CNCH_3)_6]Br_2$
    hexakis(methyl isocyanide)iron(II) bromide
11. $[Ru(HSO_3)_2(NH_3)_4]$
    tetraamminebis(hydrogensulfito)ruthenium(II)
12. $[Co(H_2O)_2(NH_3)_4]Cl_3$
    tetraamminediaquacobalt(III) chloride

Note 10c. A. Stock, *Z. angew. Chem.*, **32**, 373 (1919).
Note 10d. R V. G. Ewens and H. Bassett, *Chem. Ind.*, **27**, 131 (1949).

13. $[PtCl_2(C_5H_5N)(NH_3)]$
    amminedichloro(pyridine)platinum(II)
14. $Ba[BrF_4]_2$
    barium tetrafluorobromate(III)
15. $K[CrF_4O]$
    potassium tetrafluorooxochromate(v)
16. $[Ni(H_2O)_2(NH_3)_4]SO_4$
    tetraamminediaquanickel(II) sulfate

I-10.4.5   Names of ligands

I-10.4.5.1   *General*

The names for anionic ligands, whether inorganic or organic, end in -o. In general, if the anion name ends in -ide, -ite, or -ate, the final e is replaced by -o, giving -ido, -ito, and -ato, respectively. Enclosing marks are required for inorganic anionic ligands containing numerical prefixes, such as (triphosphato), and for thio-, seleno-, and telluro-analogues of oxoanions containing more than one oxygen atom, such as (thiosulfato-). Names of organic anions acting as ligands may be derived in a similar way. Neutral and cationic ligand names are used without modification and, except for aqua, ammine, carbonyl and nitrosyl, are placed within enclosing marks. All names listed as systematic in Tables I-10.1 to I-10.5 are IUPAC-approved.

Examples:
  1. CH_3COO^- acetato or ethanoato
  2. CH_3OSOO^- methyl sulfito
  3. $(CH_3)_2N^-$ dimethylamido
  4. CH_3CONH^- acetamido (Note 10e) or acetylamido

I-10.4.5.2   *Hydrogen as a ligand*

In its complexes hydrogen is always treated as anionic. The names of the other hydrogen isotopes are discussed in Isotopically Modified Compounds, *Pure Appl. Chem.*, **53**, 1887 (1981). Both 'hydrido' and 'hydro' are used for coordinated hydrogen, but the latter term is restricted to boron nomenclature (see Chapter I-11).

Examples:
  1. H^- hydrido
  2. D^- $[^2H]$hydrido

I-10.4.5.3   *Ligands based on halogen elements (elements of Group 17)*

The names of simple halogen anions are discussed in Section I-8.3.2. Variations allowed in coordination nomenclature are given in Section I-8.4.2.5. Table I-10.1 lists the names of some halogen derivatives encountered in coordination chemistry.

Note 10e. This common usage is not derived from the name acetamide but may be considered to be a contraction of acetylamido.

Table I-10.1 Names of some halogen-based ligands[a]

Formula	Systematic ligand name	Alternative ligand name
Br_2	(dibromine)	
F^-		fluoro
Cl^-		chloro
$(I_3)^-$		[triiodo(1$-$)]
$[ClF_2]^-$	[difluorochlorato(1$-$)]	
$[IF_4]^-$	[tetrafluoroiodato(1$-$)]	
$[IF_6]^-$	[hexafluoroiodato(1$-$)]	
$(ClO)^-$	[oxochlorato(1$-$)]	hypochlorito
$(ClO_2)^-$	[dioxochlorato(1$-$)]	chlorito
$(ClO_3)^-$	[trioxochlorato(1$-$)]	chlorato
$(ClO_4)^-$	[tetraoxochlorato(1$-$)]	perchlorato
$(IO_5)^{3-}$	[pentaoxoiodato(3$-$)]	mesoperiodato[b]
$(IO_6)^{5-}$	[hexaoxoiodato(5$-$)]	orthoperiodato[b]
$(I_2O_9)^{4-}$	[μ-oxo-octaoxodiiodato(4$-$)]	(dimesoperiodato)[b]

[a]The use of enclosing marks in these Tables is to some extent arbitrary. For example, $[IF_4]^-$ is regarded as a coordination complex, hence the square brackets. $(ClO_3)^-$ is regarded as a simple anion (Section I-8.3.3), hence the parentheses. The braces surrounding systematic names of the ligands are necessary when they are combined into the name of a coordination entity. For organic anions, the sequence of enclosing marks accords with normal organic practice, which differs from the sequence used in coordination nomenclature.

Enclosing marks are placed on ligand names in Tables I-10.1 to I-10.4 as they are to be used in names of coordination entities (see Section I-2.2.1).

[b]These traditional names are not recommended (see Chapter I-9).

I-10.4.5.4   *Ligands based on chalcogen elements* (*elements of Group 16*)

The Table of chalcogen derivatives (Table I-10.2) includes certain anions that have been assigned names that may differ from the simple rule of Section I-10.4.5.1. IUPAC-approved names are indicated with *. Those used since the 9th Collective Index of

Table I-10.2 Names for chalcogen-based ligands[a,b]

Formula	Systematic ligand name	Alternative ligand name
O_2	(dioxygen)	oxygen
S_8	(octasulfur)	
O^{2-}	oxido	oxo*[†]
S^{2-}	sulfido	thio*, thioxo[†]
Se^{2-}	selenido	selenoxo[†]
Te^{2-}	tellurido	telluroxo[†]
$(O_2)^{2-}$	[dioxido(2$-$)]	peroxo*, peroxy[†]
$(O_2)^-$	[dioxido(1$-$)]	hyperoxo*, superoxido*[†]
$(O_3)^-$	[trioxido(1$-$)]	ozonido*
$(S_2)^{2-}$	[disulfido(2$-$)]	(dithio)[†]
$(S_5)^{2-}$	[pentasulfido(2$-$)] (pentasulfane-1,5-diido)	
$(Se_2)^{2-}$	[diselenido(2$-$)]	(diseleno)[†]
$(Te_2)^{2-}$	[ditellurido(2$-$)]	(ditelluro)[†]

Table I-10.2 (*continued*)

Formula	Systematic ligand name	Alternative ligand name
H_2O		aqua*[†]
H_2S	(sulfane)	(hydrogen sulfide)*
H_2Se	(selane)	(hydrogen selenide)*
H_2Te	(tellane)	(hydrogen telluride)*
$(OH)^-$	hydroxido	hydroxo*, hydroxy[†]
$(SH)^-$	sulfanido, (hydrogensulfido)	mercapto[c]
$(SeH)^-$	selanido, (hydrogenselenido)	selenyl[†]
$(TeH)^-$	tellanido, (hydrogentellurido)	telluryl[†]
H_2O_2		(hydrogen peroxide)
H_2S_2	(disulfane)	(hydrogen disulfide)
H_2Se_2	(diselane)	(hydrogen diselenide)
H_2S_5	(pentasulfane)	(hydrogen pentasulfide)
$(HO_2)^-$		(hydrogenperoxo)*, (hydroperoxy)[†]
$(HS_2)^-$	(disulfanido)	(hydrogendisulfido), (hydrodisulfido)[†]
$(HS_5)^-$	(pentasulfanido)	(hydrogenpentasulfido)
$(CH_3O)^-$	(methanolato)	methoxo*, methoxy[†]
$(C_2H_5O)^-$	(ethanolato)	ethoxo*, ethoxy[†]
$(C_3H_7O)^-$	(1-propanolato)	propoxido*, propoxy[†]
$(C_4H_9O)^-$	(1-butanolato)	butoxido*, butoxy[†]
$(C_5H_{11}O)^-$	(1-pentanolato)	(pentyloxido)*, pentoxy[†]
$(C_{12}H_{25}O)^-$	(1-dodecanolato)	(dodecyloxido)*, (dodecyloxy)[††]
$(CH_3S)^-$	(methanethiolato)	(methylthio)*
$(C_2H_5S)^-$	(ethanethiolato)	
$(C_2H_4ClO)^-$	(2-chloroethanolato)	
$(C_6H_5O)^-$	(phenolato)	phenoxido*, phenoxy[†]
$(C_6H_5S)^-$	(benzenethiolato)	(phenylthio)*
$[C_6H_4(NO_2)O]^-$	(4-nitrophenolato)	
CO	(carbon monoxide)	carbonyl*[†]
CS	(carbon monosulfide)	(thiocarbonyl)*, (carbonothioyl)[†]
$(C_2O_4)^{2-}$	(ethanedioato)	(oxalato)[††]
$(HCO_2)^-$	(methanoato)	(formato)[††]
$(CH_3CO_2)^-$	(ethanoato)	(acetato)[††]
$(CH_3CH_2CO_2)^-$	(propanoato)	(propionato)[††]
$(SO_2)^{2-}$	[dioxosulfato(2−)]	[sulfoxylato(2−)][††]
$(SO_3)^{2-}$	[trioxosulfato(2−)]	[sulfito(2−)]
$(HSO_3)^-$	[hydrogentrioxosulfato(1−)]	(hydrogensulfito)
$(SeO_2)^{2-}$	[dioxoselenato(2−)]	[selenoxylato(2−)]
$(S_2O_2)^{2-}$	[dioxothiosulfato(2−)]	[thiosulfito(2−)][††]
$(S_2O_3)^{2-}$	[trioxothiosulfato(2−)]	[thiosulfato(2−)]
$(SO_4)^{2-}$	[tetraoxosulfato(2−)]	[sulfato(2−)]
$(S_2O_6)^{2-}$	[hexaoxodisulfato(S—S)(2−)]	[dithionato(2−)]
$(S_2O_7)^{2-}$	[μ-oxo-hexaoxodisulfato(2−)]	[disulfato(2−)]
$(TeO_6)^{6-}$	[hexaoxotellurato(6−)]	orthotellurato

[a] See footnote 'a' of Table I-10.1.
[b] See text for definition of *, [†], and [††].
[c] This name is no longer recommended for a ligand.

155

Chemical Abstracts are indicated with[†]. IUPAC alternative names marked with [††] are normally preferred in organic nomenclature. (See also footnote a of Table I-10.1)

I-10.4.5.5   *Ligands based on Group 15 elements*

The ligands of Group 15 of the Periodic Table (see Section I-3.8.1 and Table I-3.2) are often complicated to name. In particular, the hydrogen-containing ligands can give rise to

Table I-10.3 Names for derivatives of Group 15 elements as ligands[a]

Formula	Systematic name	Alternative ligand name
N_2	(dinitrogen)	
P_4	(tetraphosphorus)	
As_4	(tetraarsenic)	
N^{3-}	nitrido	
P^{3-}	phosphido	
As^{3-}	arsenido	
$(N_2)^{2-}$	[dinitrido(2−)]	
$(N_2)^{4-}$	[dinitrido(4−)]	[hydrazido(4−)]
$(N_3)^-$	(trinitrido)	azido*
$(P_2)^{2-}$	[diphosphido(2−)]	
$(CN)^-$	cyano	
$(NCO)^-$	(cyanato)	
$(NCS)^-$	(thiocyanato)	
$(NCSe)^-$	(selenocyanato)	
$(NCN)^{2-}$	[carbodiimidato(2−)]	
NF_3	(trifluoroazane)	(nitrogen trifluoride)
NH_3	(azane)	ammine*[†]
PH_3	(phosphane)	(phosphine)
AsH_3	(arsane)	(arsine)
SbH_3	(stibane)	(stibine)
$(NH)^{2-}$	azanediido	imido*[†]
$(NH_2)^-$	azanido	amido*[†]
$(PH)^{2-}$	phosphanediido	phosphinidene[†]
$(PH_2)^-$	phosphanido	phosphino[†]
$(SbH)^{2-}$	stibanediido	stibylene[†]
$(SbH_2)^-$	stibanido	stibino[†]
$(AsH)^{2-}$	arsanediido	arsinidene[†]
$(AsH_2)^-$	arsanido	arsino[†]
$(FN)^{2-}$	(fluoroazanediido)	(fluorimido)
$(ClHN)^-$	(chloroazanido)	(chloramido)
$(Cl_2N)^-$	(dichloroazanido)	(dichloramido)
$(FP)^{2-}$	(fluorophosphanediido)	
$(F_2P)^-$	(difluorophosphanido)	(difluorophosphido)
		(phosphonous difluoridato)[†]
CH_3NH_2	(methanamine)	(methylamine)[††]
$(CH_3)_2NH$	(*N*-methylmethanamine)	(dimethylamine)[††]
$(CH_3)_3N$	(*N,N*-dimethylmethanamine)	(trimethylamine)[††]
CH_3PH_2	(methylphosphane)	(methylphosphine)
$(CH_3)_2PH$	(dimethylphosphane)	(dimethylphosphine)
$(CH_3)_3P$	(trimethylphosphane)	(trimethylphosphine)
$(CH_3N)^{2-}$	[methanaminato(2−)]	(methylimido)
$(CH_3NH)^-$	[methanaminato(1−)]	(methylamido)
$[(CH_3)_2N]^-$	(*N*-methylmethanaminato)	(dimethylamido)
$[(CH_3)_2P]^-$	(dimethylphosphanido)	(dimethylphosphino)

Table I-10.3 (*continued*)

Formula	Systematic name	Alternative ligand name
$(CH_3P)^{2-}$	(methylphosphanediido)	(methylphosphinidene)[†]
$(CH_3PH)^-$	(methylphosphanido)	(methylphosphino)
$HN{=}NH$	(diazene)	(diimide), (diimine)
H_2NNH_2	(diazane)	(hydrazine)
HN_3	(hydrogen trinitride)	(hydrogen azide)*
$(HN{=}N)^-$	(diazenido)	(diiminido)
$(HNN)^{3-}$	(diazanetriido)	[hydrazido(3−)]
$(H_2NN)^{2-}$	(diazane-1,1-diido)	[hydrazido(2−)-N,N]
$(HN{-}NH)^{2-}$	(diazane-1,2-diido)	[hydrazido(2−)-N,N']
$(H_2N{-}NH)^-$	(diazanido)	(hydrazido)
$HP{=}PH$	(diphosphene)	
$H_2P{-}PH_2$	(diphosphane)	
$(HP{=}P)^-$	(diphosphenido)	
$(H_2P{-}P)^{2-}$	(diphosphane-1,1-diido)	
$(HP{-}PH)^{2-}$	(diphosphane-1,2-diido)	
$(H_2PPH)^-$	(diphosphanido)	
$HAs{=}AsH$	(diarsene)	
H_2AsAsH_2	(diarsane)	
$(HAsAs)^{3-}$	(diarsanetriido)	
$(H_2AsAs)^{2-}$	(diarsane-1,1-diido)	
$(CH_3AsH)^-$	(methylarsanido)	(methylarsino)[†]
$(CH_3As)^{2-}$	(methylarsanediido)	(methylarsinidene)[†]
H_2NOH	(hydroxyazane)	(hydroxylamine)
$(HNOH)^-$	(hydroxylaminato-κN)	(hydroxylamido)[b]
$(H_2NO)^-$	(hydroxylaminato-κO)	(hydroxylamido)[b]
$(HNO)^{2-}$	[hydroxylaminato(2−)]	(hydroxylimido)[b]
$(PO_3)^{3-}$	[trioxophosphato(3−)]	[phosphito(3−)]
$(HPO_2)^{2-}$	[hydridodioxophosphato(2−)]	[phosphonito(2−)]
$(H_2PO)^-$	[dihydridooxophosphato(1−)]	(phosphonito)
$(AsO_3)^{3-}$	[trioxoarsenato(3−)]	[arsenito(3−)]
$(HAsO_2)^{2-}$	[hydridodioxoarsenato(2−)]	[arsenito(2−)]
$(H_2AsO)^-$	[dihydridooxoarsenato(1−)]	(arsinito)
$(PO_4)^{3-}$	[tetraoxophosphato(3−)]	[phosphato(3−)]
$(HPO_3)^{2-}$	[hydridotrioxophosphato(2−)]	[phosphonato(2−)]
$(H_2PO_2)^-$	[dihydridodioxophosphato(1−)]	(phosphinato)
$(AsO_4)^{3-}$	[tetraoxoarsenato(3−)]	[arsenato(3−)]
$(HAsO_3)^{2-}$	[hydridotrioxoarsenato(2−)]	[arsonato(2−)]
$(H_2AsO_2)^-$	[dihydridodioxoarsenato(1−)]	arsinato
$(P_2O_7)^{4-}$	[μ-oxo-hexaoxodiphosphato(4−)]	[diphosphato(4−)]
$(C_6H_5N_2)^-$	(phenyldiazenido)	(phenylazo)[††]
$(NO_2)^-$	[dioxonitrato(1−)]	nitrito-O
	[dioxonitrato(1−)]	nitrito-N, nitro
$(NO_3)^-$	[trioxonitrato(1−)]	nitrato
NO	(nitrogen monoxide)	nitrosyl
NS	(nitrogen monosulfide)	(thionitrosyl)
N_2O	(dinitrogen oxide)	
$(N_2O_2)^{2-}$	[dioxodinitrato($N{-}N$)(2−)]	hyponitrito

[a] See text for definition of *, [†], and [††] and footnote 'a' of Table I-10.1.
[b] This is a broadly used trivial name which in substitutive nomenclature should be hydroxy-amido. Neither the traditional nor the systematic ligand name, (hydroxylamido) and (hydroxyl-imido), indicates specific charge location or implies bonding isomerism in a compound name.

derivatives from which one or more hydrogen atoms have been lost. The conventions used in Table I-10.3 are those already indicated above (Section I-10.4.5.4 and Table I-10.1). Ligands such as NCS– that can bond through either of two or more atoms are termed ambidentate and their linkage isomerism is treated in Sections I-10.6.2.1 and I-10.6.2.2.

I-10.4.5.6   *Organic ligands*

Names of neutral organic compounds are used as ligand names without modification. The names are modified if the ligand is charged. These ligand names should be assigned in accordance with the *Nomenclature of Organic Chemistry*, 1979 edition. For ligands containing ring systems, the *Ring Systems Handbook*, published by the American Chemical Society, Columbus, Ohio, 1984, provides a primary reference. Older forms of organic ligand name are in common usage and some of these which are frequently encountered appear in Table I-10.4. It should be stressed that IUPAC names are always preferred to trivial variants which are not self-explanatory and which have often been used in different ways by different authors. The conventions used here are presented in Section I-10.4.5.1 and Table I-10.1.

   Names of ligands derived from neutral organic compounds by the formal loss of hydrons (other than those named in Sections I-10.4.5.1 and I-8.3) are given the ending -ato. Enclosing marks are used to set off all such organic ligand names, regardless of whether they are neutral, charged, or are substituted or unsubstituted, e.g., (benzaldehyde), (benzoato), (*p*-chlorophenolato), [(2-(chloromethyl)-1-naphtholato]. The name of a cation when it is coordinated is used without change. Ligands binding to metals through carbon atoms are treated in the section on organometallic compounds (see Section I-10.9). These ligands are given radical names (see Examples in Section I-7.3.3). Common hydrocarbon radical names (see Section I-10.9) need not be set off with parentheses.

I-10.4.5.7   *Use of abbreviation*

Abbreviations are commonly used in the chemical literature. A list of ligand abbreviations is given in Table I-10.5. The following practices should be followed in the use of abbreviations. It should be assumed that the reader will not be familiar with the abbreviations. Consequently, all text should explain the abbreviations it uses. Abbreviations should be as short as possible. Abbreviations should not cause confusion. For example, the commonly accepted abbreviations for organic groups (Me, methyl; Et, ethyl; Ph, phenyl, etc.) should not be used with any other meanings. The most useful abbreviations are those that readily suggest the ligand in question, either because they are obviously derived from the ligand names or because they are systematically related to structure. A genuine effort on the part of various research groups in· a given area to standardize their use of abbreviations is essential to communication. The use of local, trivial names is discouraged. The sequential positions of ligand abbreviations in formulae should be in accord with Chapter I-4. Lower case letters are recommended for all abbreviations, except for certain hydrocarbon radicals. In formulae, the ligand abbreviation should be set off with parentheses, as in $[Co(en)_3]^{3+}$. In Table I-10.5, those hydrogen atoms that can be replaced by the metal atom are shown in the abbreviation by the symbol H. Thus, the molecule Hacac forms an anionic ligand that is abbreviated acac.

Table I-10.4 Examples of organic ligand names[a]

Systematic name	Alternative name
(1,2-ethanediamine)	(ethylenediamine)[††]
(1,2-propanediamine)	(propylenediamine)[††]
(1,3-propanediamine)	(trimethylenediamine)[††]
[N-(2-aminoethyl)-1,2-ethanediamine]	(diethylenetriamine)
[N,N'-bis(2-aminoethyl)-1,2-ethanediamine]	(triethylenetetramine)
[N,N-bis(2-aminoethyl)-1,2-ethanediamine]	[tris(2-aminoethyl)amine][††]
[N,N-bis[2-(dimethylamino)ethyl]-N',N'-dimethyl-1,2-ethanediamine]	tris[2-(dimethylamino)ethyl]amine[††]
(2-aminoethanol)	(ethanolamine)
(2,2',2''-nitrilotriethanol)	(triethanolamine)
(2,2'-bipyridine)	
(2,2':6',2''-terpyridine)	
(2,4-pentanedionato)	(acetylacetonato)*
(2,3-butanedione dioximato)	(dimethylglyoximato)
(8-quinolinolato)	(8-hydroxyquinolinato)[b]
(2-hydroxybenzaldehydato)	(salicylaldehydato)*
(1,5-diphenylthiocarbazonato)	(dithizonato), (phenyldiazenecarbothioic acid 2-phenylhydrazidato)*
N-nitroso-N-phenylhydroxylamine ammonium salt	cupferron
[1,2-ethanediylbis(dimethylphosphine)]	[1,2-bis(dimethylphosphino)ethane]
[iminodiacetato(2−)]	[N-(carboxymethyl)glycinato(2−)]*
[nitrilotriacetato(3−)]	[N,N'-bis(carboxymethyl)glycinato(3−)]*
[(1,2-ethanediyldinitrilo)tetraacetato(4−)]	[ethylenediaminetetraacetato(4−)]*
[2,2'-[1,2-ethanediylbis(nitrilomethylidyne)]-diphenolato(2−)]	[N,N'-ethylenebis(salicylideneiminato)(2−)]*, [bis(salicylal)ethylenediaminato(2−)]
(1,4,8,11-tetraazacyclotetradecane)	
(1,4,7,10,13,16-hexaoxacyclooctadecane)	
(N,N-diethylcarbamodithioato)	(diethyldithiocarbamato)*
(O-ethyl carbonodithioato)	(O-ethyl dithiocarbonato)*, (ethyl xanthato)
[3,7,12,17-tetramethyl-8,13-divinyl-porphyrin-2,18-dipropionato(2−)]	[protoporphyrin IXato(2−)]

[a]See text for definition of *, †, and ††.
[b]Oxine, a name for pyran derived from the Hantzsch–Widman system, is not recommended.

I-10.5   STEREOCHEMICAL DESCRIPTORS

I-10.5.1   **General**

Different geometrical arrangements of the atoms attached to the central atom are possible for all coordination numbers greater than one. Thus, two-coordinate species may involve a linear or a bent disposition of the ligands and central atom. Similarly, three-coordinate species may be trigonal planar or trigonal pyramidal and four-coordinate species may be square planar, square pyramidal, or tetrahedral. The coordination polyhedron (or polygon in planar molecules) may be denoted in the name by an affix

Table I-10.5 Abbreviations for ligands and ligand forming compounds

Abbreviation	Common name	Systematic name
Diketones		
Hacac	acetylacetone	2,4-pentanedione
Hhfa	hexafluoroacetylacetone	1,1,1,5,5,5-hexafluoro-2,4-pentanedione
Hba	benzoylacetone	1-phenyl-1,3-butanedione
Hfod	1,1,1,2,2,3,3-heptafluoro-7,7-dimethyl-4,6-octanedione	6,6,7,7,8,8,8-heptafluoro-2,2-dimethyl-3,5-octanedione
Hfta	trifluoroacetylacetone	1,1,1-trifluoro-2,4-pentanedione
Hdbm	dibenzoylmethane	1,3-diphenyl-1,3-propanedione
Hdpm	dipivaloylmethane	2,2,6,6-tetramethyl-3,5-heptanedione
Amino alcohols		
Hea	ethanolamine	2-aminoethanol
H₃tea	triethanolamine	2,2′,2″-nitrilotriethanol
H₂dea	diethanolamine	2,2′-iminodiethanol
Hydrocarbons		
cod	cyclooctadiene	1,5-cyclooctadiene
cot	cyclooctatetraene	1,3,5,7-cyclooctatetraene
Cp	cyclopentadienyl	cyclopentadienyl
Cy	cyclohexyl	cyclohexyl
Ac	acetyl	acetyl
Bu	butyl	butyl
Bzl	benzyl	benzyl
Et	ethyl	ethyl
Me	methyl	methyl
nbd	norbornadiene	bicyclo[2.2.1]hepta-2,5-diene
Ph	phenyl	phenyl
Pr	propyl	propyl

Heterocycles

py	pyridine	pyridine
thf	tetrahydrofuran	tetrahydrofuran
Hpz	pyrazole	1H-pyrazole
Him	imidazole	1H-imidazole
terpy	2,2':2''-terpyridine	2,2':6',2''-terpyridine
picoline	α-picoline	2-methylpyridine
Hbpz$_4$	hydrogen tetra(1-pyrazolyl)borate(1−)	hydrogen tetrakis(1H-pyrazolato-N)borate(1−)
isn	isonicotinamide	4-pyridinecarboxamide
nia	nicotinamide	3-pyridinecarboxamide
pip	piperidine	piperidine
lut	lutidine	2,6-dimethylpyridine
Hbim	benzimidazole	1H-benzimidazole

Chelating and other ligands

H$_4$edta	ethylenediaminetetraacetic acid	(1,2-ethanediyldinitrilo)tetraacetic acid
H$_5$dtpa	N,N,N',N'',N''-diethylenetriaminepentaacetic acid	[[(carboxymethyl)imino]bis(ethylenenitrilo)]tetraacetic acid
H$_3$nta	nitrilotriacetic acid	
H$_4$cdta	trans-1,2-cyclohexanediaminetetraacetic acid	trans-(1,2-cyclohexanediyldinitrilo)tetraacetic acid
H$_2$ida	iminodiacetic acid	
dien	diethylenetriamine	N-(2-aminoethyl)-1,2-ethanediamine
en	ethylenediamine	1,2-ethanediamine
pn	propylenediamine	1,2-propanediamine
tmen	N,N,N',N'-tetramethylethylenediamine	N,N,N',N'-tetramethyl-1,2-ethanediamine
tn	trimethylenediamine	1,3-propanediamine
tren	tris(2-aminoethyl)amine	tris(2-aminoethyl)amine
trien	triethylenetetramine	N,N'-bis(2-aminoethyl)-1,2-ethanediamine
chxn	1,2-diaminocyclohexane	1,2-cyclohexanediamine
hmta	hexamethylenetetramine	1,3,5,7-tetraazatricyclo[3.3.1.1^{3,7}]decane
Hthsc	thiosemicarbazide	hydrazinecarbothioamide†
depe	1,2-bis(diethylphosphino)ethane	1,2-ethanediylbis(diethylphosphine)

Table I-10.5 (*continued*)

Abbreviation	Common name	Systematic name
diars	o-phenylenebis(dimethylarsine)	1,2-phenylenebis(dimethylarsine)
dppe	1,2-bis(diphenylphosphino)ethane	1,2-ethanediylbis(diphenylphosphine)
diop	2,3-O-isopropylidene-2,3-dihydroxy-1,4-bis(diphenylphosphino)butane	3,4-bis[(diphenylphosphinyl)methyl]-2,2-dimethyl-1,3-dioxolane
triphos		[2-[(diphenylphosphino)methyl]-2-methyl-1,3-propanediyl]bis(diphenylphosphine)
hmpa	hexamethylphosphoric triamide	hexamethylphosphoric triamide
bpy	2,2'-bipyridine	2,2'-bipyridine
H₂dmg	dimethylglyoxime	2,3-butanedione dioxime
dmso	dimethyl sulfoxide	sulfinyldimethane
phen	1,10-phenanthroline	1,10-phenanthroline
tu	thiourea	thiourea
Hbig	biguanide	imidodicarbonimidic diamide
HEt₂dtc	diethyldithiocarbamic acid	diethylcarbamodithioic acid
H₂mnt	maleonitriledithiol	2,3-dimercapto-2-butenedinitrile
tcne	tetracyanoethylene	ethenetetracarbonitrile
tcnq	tetracyanoquinodimethane	2,2'-(2,5-cyclohexadiene-1,4-diylidene)bis(1,3-propanedinitrile)
dabco	triethylenediamine	1,4-diazabicyclo[2.2.2]octane
2,3,2-tet	1,4,8,11-tetraazaundecane	N,N'-bis(2-aminoethyl)-1,3-propanediamine
3,3,3-tet	1,5,9,13-tetraazatridecane	N,N'-bis(3-aminopropyl)-1,3-propanediamine
ur	urea	urea
dmf	dimethylformamide	N,N-dimethylformamide
Schiff base		
H₂salen	bis(salicylidene)ethylenediamine	2,2'-[1,2-ethanediylbis(nitrilomethylidyne)]diphenol
H₂acacen	bis(acetylacetone)ethylenediamine	4,4'-(1,2-ethanediyldinitrilo)bis(2-pentanone)

H_2salgly	salicylideneglycine	N-[(2-hydroxyphenyl)methylene]glycine
H_2saltn	bis(salicylidene)-1,3-diaminopropane	2,2′-[1,3-propanediylbis(nitrilomethylidyne)]-diphenol
H_2saldien	bis(salicylidene)diethylenetriamine	2,2′-[iminobis(1,2-ethanediylnitrilomethylidyne)]diphenol
H_2tsalen	bis(2-mercaptobenzylidene)ethylenediamine	2,2′-[1,2-ethanediylbis(nitrilomethylidyne)]dibenzenethiol
Macrocycles		
18-crown-6	1,4,7,10,13,16-hexaoxacyclooctadecane	1,4,7,10,13,16-hexaoxacyclooctadecane
benzo-15-crown-5	2,3-benzo-1,4,7,10,13-pentaoxacyclopentadec-2-ene	2,3,5,6,8,9,11,12-octahydro-1,4,7,10,13-benzopentaoxacyclopentadecene
cryptand 222	4,7,13,16,21,24-hexaoxa-1,10-diazabicyclo[8.8.8]hexacosane	4,7,13,16,21,24-hexaoxa-1,10-diazabicyclo[8.8.8]hexacosane
cryptand 211	4,7,13,18-tetraoxa-1,10-diazabicyclo[8.5.5]icosane	4,7,13,18-tetraoxa-1,10-diazabicyclo[8.5.5]icosane
[12]aneS_4	1,4,7,10-tetrathiacyclododecane	1,4,7,10-tetrathiacyclododecane
H_2pc	phthalocyanine	phthalocyanine
H_2tpp	tetraphenylporphyrin	5,10,15,20-tetraphenylporphyrin
H_2oep	octaethylporphyrin	2,3,7,8,12,13,17,18-octaethylporphyrin
ppIX	protoporphyrin IX	3,7,12,17-tetramethyl-8,13-divinylporphyrin-2,18-dipropanoic acid
[18]aneP_4O_2	1,10-dioxa-4,7,13,16-tetraphosphacyclooctadecane	1,10-dioxa-4,7,13,16-tetraphosphacyclooctadecane
[14]aneN_4	1,4,8,11-tetraazacyclotetradecane	1,4,8,11-tetraazacyclotetradecane
[14]1,3-dieneN_4	1,4,8,11-tetraazacyclotetradeca-1,3-diene	1,4,8,11-tetraazacyclotetradeca-1,3-diene
Me_4[14]-aneN_4	2,3,9,10-tetramethyl-1,4,8,11-tetraazacyclotetradecane	2,3,9,10-tetramethyl-1,4,8,11-tetraazacyclotetradecane
cyclam	1,4,8,11-tetraazacyclotetradecane	1,4,8,11-tetraazacyclotetradecane

called the *polyhedral symbol*. This descriptor clearly distinguishes isomers differing in the geometries of their coordination polyhedra.

Given the same coordination polyhedron, diastereoisomerism can arise when the ligands are not all alike as with the *cis* and *trans* isomers of tetraamminedichlorochromium(III), diamminedichloroplatinum(II), and bis(2-aminoethanethiolato)nickel(II) (Examples 1–6). Attempts to produce descriptors similar to *cis* and *trans* for stereochemically more complicated coordination entities have failed to achieve generality, and labels such as *fac* and *mer* are no longer recommended. Nevertheless, a diastereoisomeric structure may be indicated for any polyhedron using a *configuration index* as an affix to the name or formula.

Finally, the chiralities of enantiomeric structures can be indicated using *chirality symbols*. The first two indexes are presented in this Section, while chirality is discussed in Section I-10.7.

Examples:

1. *trans*-isomer   2. *cis*-isomer

3. *cis*-isomer   4. *trans*-isomer

5. *trans*-isomer   6. *cis*-isomer

I-10.5.2   Polyhedral symbol

The polyhedral symbol indicates the geometrical arrangements of the coordinating atoms about the central atom. This symbol must be assigned before any other stereochemical features can be considered. It consists of one or more capital italic letters derived from common geometric terms which denote the idealized geometry of the ligands around the coordination centre, and an arabic numeral that is the coordination number of the central atom. The polyhedral symbol is used as an affix, enclosed in parentheses, and separated from the name by a hyphen. The polyhedral symbols for the most common coordination geometries for coordination numbers 2 to 9 inclusive are given in Table I-10.6, and the corresponding structures are shown in Table I-10.7.

Table I-10.6 List of polyhedral symbols[a]

Coordination polyhedron	Coordination number	Polyhedral symbol
linear	2	L-2
angular	2	A-2
trigonal plane	3	TP-3
trigonal pyramid	3	TPY-3
tetrahedron	4	T-4
square plane	4	SP-4
square pyramid	4	SPY-4
trigonal bipyramid	5	TBPY-5
square pyramid	5	SPY-5
octahedron	6	OC-6
trigonal prism	6	TPR-6
pentagonal bipyramid	7	PBPY-7
octahedron, face monocapped	7	OCF-7
trigonal prism, square face monocapped	7	TPRS-7
cube	8	CU-8
square antiprism	8	SAPR-8
dodecahedron	8	DD-8
hexagonal bipyramid	8	HBPY-8
octahedron, trans-bicapped	8	OCT-8
trigonal prism, triangular face bicapped	8	TPRT-8
trigonal prism, square face bicapped	8	TPRS-8
trigonal prism, square face tricapped	9	TPRS-9
heptagonal bipyramid	9	HBPY-9

[a]Strictly, not all the geometries can be represented by polyhedra.

Table I-10.7 Polyhedral symbols and geometrical structures

Polyhedra of four-coordination

tetrahedron   square planar

T-4   SP-4

Table I-10.7 Polyhedral symbols (*continued*)

Polyhedra of five-coordination

trigonal bipyramid   square pyramid

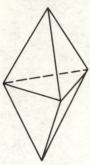

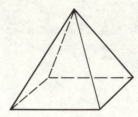

TBPY-5   *SPY*-5

Polyhedra of six-coordination

octahedron   trigonal prism

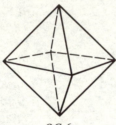

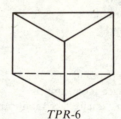

OC-6   *TPR*-6

Polyhedra of seven-coordination

pentagonal   octahedron, face   trigonal prism,
bipyramid   monocapped   square face
                                              monocapped

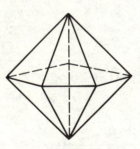

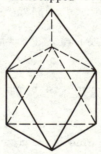

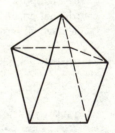

PBPY-7   *OCF*-7   *TPRS*-7

Polyhedra of eight-coordination

         square   hexagonal
cube   antiprism   dodecahedron   bipyramid

CU-8   *SAPR*-8   *DD*-8   *HBPY*-8

Table I-10.7 (*continued*)

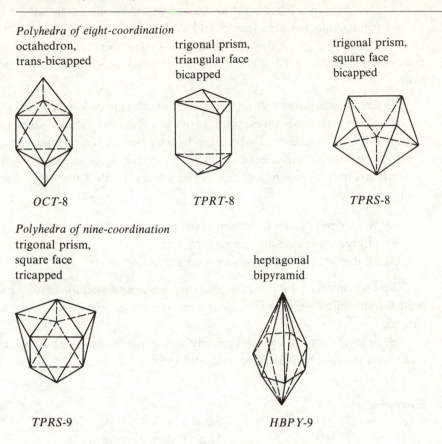

Polyhedra of eight-coordination

octahedron, trans-bicapped

trigonal prism, triangular face bicapped

trigonal prism, square face bicapped

OCT-8 | *TPRT*-8 | *TPRS*-8

Polyhedra of nine-coordination

trigonal prism, square face tricapped

heptagonal bipyramid

TPRS-9 | *HBPY*-9

Distortions from idealized geometries commonly occur. However, it is normal practice to relate molecular structures to idealized models. For real molecules, the stereochemical descriptor should be based on the closest idealized geometry. No specific criteria for making decisions are recommended, and in a few exceptional cases clear assignments may not be possible.

I-10.5.3 Configuration index

I-10.5.3.1 *Definition of index and assignment of priority numbers to ligating atoms*

Having developed descriptors for the general geometry of coordination compounds, it becomes necessary to identify the individual coordination positions. The *configuration index* is a series of digits identifying the positions of the ligating atoms on the vertices of the coordination polyhedron. The individual configuration index has the property that it distinguishes between diastereoisomers. The digits of the configuration index are established by assigning an order of priority to the ligating atoms as described in Section I-10.5.3.1. The configuration index appears within the square brackets enclosing the site symmetry symbol, following that symbol, and separated from it by a hyphen (see Examples in Section I-10.5.4).

The procedure for assigning priority numbers to the ligating atoms of a mononuclear coordination system is based on the standard sequence rule developed for enantiomeric carbon compounds by Cahn, Ingold, and Prelog [R. S. Cahn, C. Ingold, and V. Prelog, *Angew. Chem., Int. Ed. Engl.*, **5**, 385 (1966); V. Prelog and G. Helmchen, *Angew. Chem., Int. Ed. Engl.*, **21**, 567 (1982)]. These are often referred to as the CIP rules and are quoted below.

"The ligands associated with an element of chirality are ordered by comparing them at each step in bond-by-bond exploration of them, from the element, along the successive bonds of each ligand, and where the ligands branch, first along branch-pathways providing highest precedence to their respective ligands, the explorations being continued to total ordering by use of the Standard Sub-Rules each to exhaustion in turn, namely:

(0) Nearer end of axis or side of plane precedes further.
(1) Higher atomic number precedes lower.
(2) Higher atomic mass-number precedes lower."

There are further CIP sub-rules, but they are not needed for present purposes. The need for sub-rule (0) for mononuclear coordination entities has not yet been demonstrated.

The ligating atom with highest priority is assigned the priority number 1; the ligating atom with the next highest priority, 2; and so on.

Examples:

1.

Sequence sub-rule (1)
Higher atomic number
precedes lower.

Priority sequence:
$Br > Cl > PPh_3, PPh_3 > NMe_3 > CO$
Priority numbers sequence:
$1 > 2 > 3, 3 > 4 > 5$

2.

Sequence sub-rule (1)
Higher atomic number precedes
lower, exploration of ligand
structure.

In Example 3, all the ligating atoms are nitrogen atoms. The key illustrates how proceeding along the branches of the ligand constituents allows priorities to be assigned. The numbers in columns 1, 2, and 3 on the right are the atomic numbers of the atoms in the structures. The averaging techniques used in the case of resonance structures (last two ligands in the list) are given in the references cited at the beginning of Section I-10.5.3.1.

Example:
3.

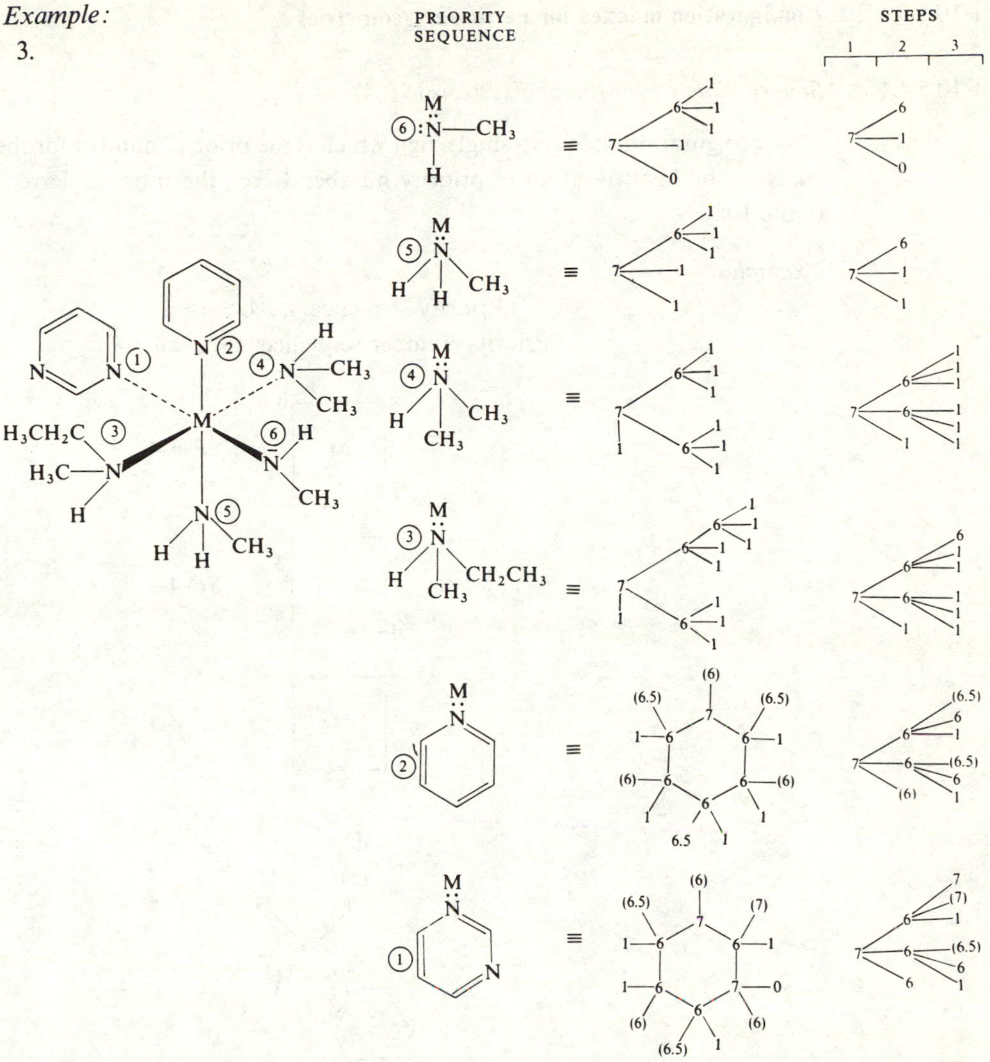

PRIORITY SEQUENCE

STEPS

I-10.5.3.2 Differentiation between ligating atoms with identical priority number

Using the sequence rule presented in Section I-10.5.3.1, priority numbers are assigned to ligating atoms on the basis of differences in chemical constitution. Often, therefore, different ligating atoms in a mononuclear coordination entity will have the same priority number. In order to use these priority numbers to provide an exact description for such coordination systems, a principle called the '*trans* maximum difference of priority numbers' is applied, when appropriate.

I-10.5.3.3 Trans *maximum difference of priority numbers* (*for coordination numbers* 4, 5, *and* 6)

When it is necessary to distinguish between ligating atoms having identical priority numbers, the ligating atom *trans* or opposite (on a structural axis) to the ligating atom of highest priority number is chosen. This, and the priming principle for chelated ligands, an additional rule to be stated in Section I-10.6.3.2, leads to unique configuration indexes that have no more than three digits for coordination numbers 4, 5, and 6.

I-10.5.4   Configuration indexes for particular geometries

I-10.5.4.1   *Square planar coordination systems (SP-4)*

The configuration index is a single digit which is the priority number for the ligating atom *trans* to the ligating atom of priority number 1, i.e., the most preferred ligating atom (Note 10f).

Examples:

Priority sequence: a > b > c > d
Priority number sequence: 1 < 2 < 3 < 4

1.

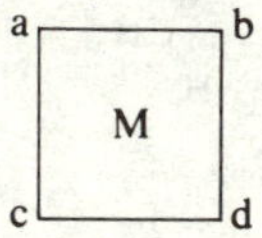

SP-4-4

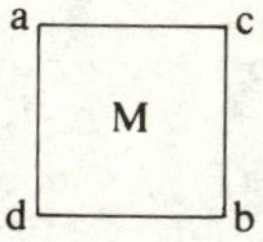

SP-4-2

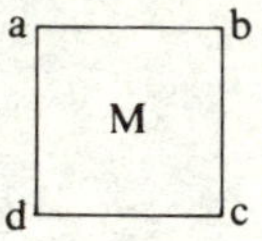

SP-4-3

2.

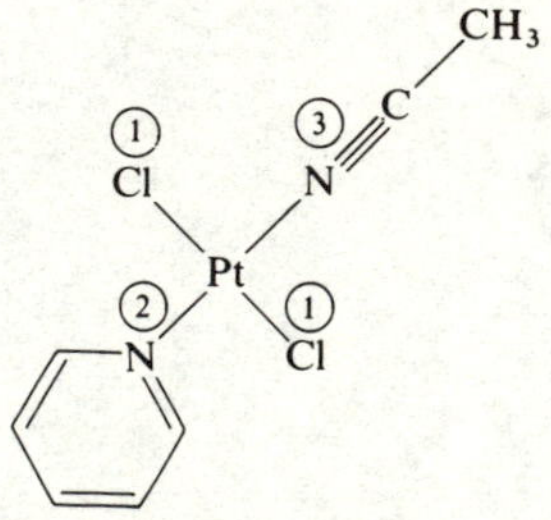

[*SP*-4-1]-(acetonitrile)dichloro(pyridine)platinum(II)

3.

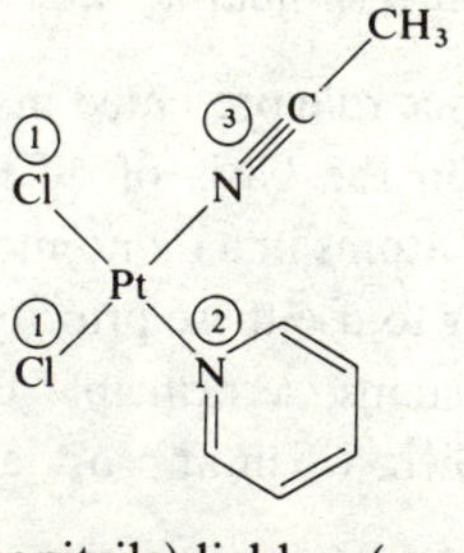

[*SP*-4-3]-(acetonitrile)dichloro(pyridine)platinum(II)

Note 10f. *cis–trans* terminology alone is not adequate to distinguish between the three isomers of a square planar coordination entity [Mabcd].

I-10.5.4.2   *Octahedral coordination systems (OC-6)*

The configuration index consists of two digits. The first digit is the priority number of the ligating atom *trans* to the ligating atom of priority number 1, i.e., the preferred ligating atom. These two ligating atoms define the reference axis of the octahedron. The second digit of the configuration index is the priority number of the ligating atom *trans* to the ligating atom with the lowest priority number (most preferred) in the plane that is perpendicular to the reference axis (the *trans* maximum principle, Section I-10.5.3.3).

Examples:

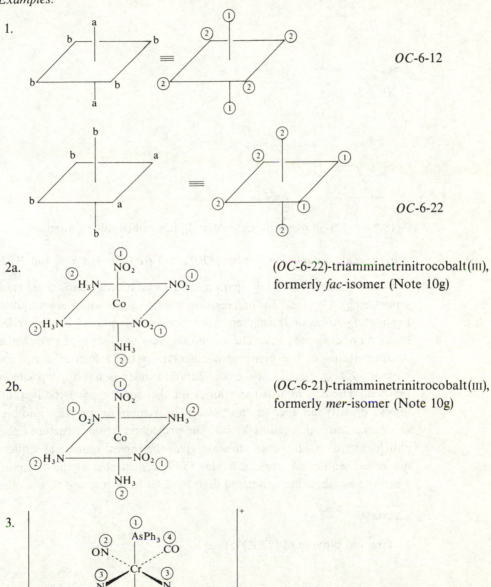

1.   *OC*-6-12

                                                                                      OC-6-22

2a.   (*OC*-6-22)-triamminetrinitrocobalt(III),
                                                              formerly *fac*-isomer (Note 10g)

2b.   (*OC*-6-21)-triamminetrinitrocobalt(III),
                                                              formerly *mer*-isomer (Note 10g)

3.

(*OC*-6-43)-bis(acetonitrile)dicarbonylnitrosyl(triphenylarsine)chromium(1 +)

Note 10g. The isomer designators *fac* and *mer* may be useful for general discussions but are not recommended for nomenclature purposes.

I-10.5.4.3   *Square pyramidal coordination systems (SPY-5)*

The configuration index consists of two digits. The first digit is the priority number of the ligating atom on the C_4 symmetry axis of the idealized pyramid. The second digit is the priority number for the ligating atom *trans* to the ligating atom with the lowest priority number in the plane perpendicular to the C_4 symmetry axis.

Examples:

1. *SPY*-5-43

2. (*SPY*-5-12)-dibromotris[di(*tert*-butyl)phenylphosphine]palladium

I-10.5.4.4   *Bipyramidal coordination systems (TBPY-5, PBPY-7, HBPY-8, and HBPY-9)*

The configuration index for bipyramidal coordination systems consists of two segments separated by a hyphen. The first segment has two digits which are the priority numbers of the ligating atoms on the highest order rotational symmetry axis, the reference axis. The lower number is cited first. The second segment consists of the priority numbers of the ligating atoms in the plane perpendicular to the reference axis. (For the trigonal bipyramid, this segment of the configuration index is not appropriate and is therefore omitted.) The first digit is the priority number for the preferred ligating atom, i.e. the lowest priority number in the plane. The remaining priority numbers are cited in sequential order proceeding around the projection of the structure either clockwise or anticlockwise, in whichever direction gives the lowest numerical sequence. The lowest numerical sequence is the one having the lower number at the first point of difference when the numbers are compared digit by digit from one end to the other.

Examples:

1. Trigonal bipyramid (*TBPY-5*)

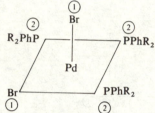

TBPY-5-25

(*TBPY*-5-11)-tricarbonylbis(triphenylphosphine)iron

172

2. Pentagonal bipyramid (*PBPY*-7)

PBPY-7-34-12342 (Note 10h)

Additional structures and further elaboration of rules appear in Section I-10.6, on stereochemistry of metal chelate complexes.

I-10.6   FORMULAE AND NAMES FOR CHELATE COMPLEXES

I-10.6.1   General

The chelating ligands are treated in accordance with the rules specified for monodentate ligands but with certain additional and necessary considerations. Specifically, coordination formulae use the same sequence of symbols and the same conventions with regard to the use of enclosing marks, ionic charges, and oxidation numbers. However, attention must be given to the use of enclosing marks within organic ligand formulae and names, since this involves the nesting sequence. Chelating ligands introduce additional questions concerning which ligand atoms are ligated to the central atom, and further stereochemical relationships.

I-10.6.2   Designation of the ligating atoms in a polydentate ligand

I-10.6.2.1   *Donor atom symbol as the index*

A polydentate ligand possesses more than one donor site, some or all of which may be involved in coordination. In earlier practice, the different donors of the ligand were denoted by adding to the end of the name of the ligand the italicized symbol(s) for the atom or atoms through which attachment to the metal occurs. Thus, dithiooxalate anion conceivably may be attached through S or O, and these were distinguished by dithiooxalato-*S,S'* and dithiooxalato-*O,O'*, respectively. The following system of citation of ligating atoms is suggested for simple cases.

For ligands with ligating atoms linearly arranged along a chain, the order of citation symbols should be successive, starting at one end. The choice of end is based upon alphabetical order if the ligating atoms are different, e.g., cysteinato-*N,S*; cysteinato-*N,O*. When no such simple distinction can be made, the ligating atom at the site with the lowest *locant* assigned according to organic practice is given preference.

Examples:

1. $(CF_3COCHCOH_3)^-$ 1,1,1-trifluoro-2,4-pentanedionato, CF_3CO preferred to CH_3CO

2. $H_2NCH(CH_3)CH_2NH_2$ 1,2-propanediamine, NH_2CH_2 preferred to $NH_2CH(CH)_3$

Note 10h. The series 12342 is lexicographically lower than 12432.

If the same element is involved in the different positions, the place in the chain or ring to which the central atom is attached is indicated by numerical superscripts. It may be necessary to provide numerical superscripts where none are available from normal organic substitutive nomenclature.

Examples:

3. $(CH_3COCHCOCH_3)$ 2,4-pentanedionato-C^3

4.

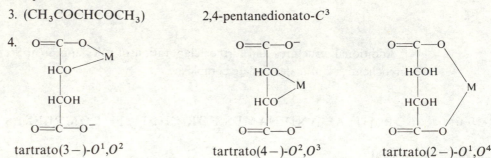

tartrato(3−)-O^1,O^2 tartrato(4−)-O^2,O^3 tartrato(2−)-O^1,O^4

I-10.6.2.2   *The kappa convention*

As the complexity of the ligand name increases, a more general system is needed to indicate the points of ligation. In the nomenclature of polydentate chelate complexes, single ligating atom attachments of a polyatomic ligand to a coordination centre are indicated by the italic element symbol preceded by a Greek kappa, κ.

Monodentate ambident ligands provide simple examples, although for these cases the kappa convention is not significantly more useful than the 'donor atom symbol' convention (Section I-10.6.2.1). Nitrogen-bonded NCS is thiocyanato-κN and sulfur-bonded NCS is thiocyanato-κS. Nitrogen-bonded nitrite is named nitrito-κN and oxygen-bonded nitrite is nitrito-κO, i.e., $[O=N-O-Co(NH_3)_5]^{2+}$ is pentaamminenitrito-κO-cobalt(III) ion. In cases where two or more such ligating groups are involved, a superscript is used on κ.

Example:

1. $[Ni\{(CH_3)_2PCH_2CH_2P(CH_3)_2\}Br_2]$
   dibromobis[1,2-ethanediylbis(dimethylphosphine)-$\kappa^2 P$]nickel(II)

In the case of more complicated ligand names, the ligand locant is placed after that portion of the ligand name which denotes the particular function, ring, chain, or radical in which the ligating atom is found. Ligating atoms occurring in functions, chains, rings, and radicals which contain other donor atoms are uniquely indicated by a superscript numeral, letter, or prime on the element symbol. These indexes denote the position of the ligating atom in the function, chain, ring, or radical. The following Example shows the result of insertion of nickel into a C–H bond. The locant of the carbon atom is indicated by a right superscript on the italicized donor atom symbol.

Example:

2.

[2(-diphenylphosphino-κP)-phenyl-κC^1]hydrido(triphenylphosphine-κP)nickel(II)

For polydentate ligands, a right superscript numeral is added to the symbol κ in order to indicate the number of identically bound ligating atoms in the *flexidentate* (Note 10i) ligand. Any doubling prefixes, such as bis-, are presumed to operate on the κ locant index as well. Thus, one uses the partial name '... bis(2-amino-κN-ethyl)...' and NOT '... bis(2-amino-$\kappa^2 N$-ethyl)...' in Example 3 below. Examples 3 and 4 use tridentate chelation by the linear tetraamine ligand, N,N'-bis(2-aminoethyl)-1,2-ethanediamine to illustrate these rules.

Examples:

3.

[N,N'-bis(2-amino-κN-ethyl)-1,2-ethanediamine-κN]chloroplatinum(II)

4.

[N-(2-amino-κN-ethyl)-N'-(2-aminoethyl)-1,2-ethanediamine-$\kappa^2 N,N'$]chloroplatinum(II)

For Example 3, coordination by the two terminal primary amino-groups of the ligand is indicated by placing the index after the prefix radical function name and the doubling is indicated by the doubling prefix bis- that precedes the phrase (2-amino-κN-ethyl). The appearance of the simple index κN after the word ethanediamine clearly indicates the binding by only one of the two equivalent secondary amino nitrogen atoms. In Example 4, only one of the primary amines is coordinated, and this is indicated by not using the doubling prefix bis- and repeating (2-aminoethyl), but inserting the κ index only in the first, i.e., (2-amino-κN-ethyl). The involvement of both of the secondary ethanediamine nitrogen atoms in chelation is indicated by the index $\kappa^2 N,N'$.

The unambiguous tridentate chelation by the tetrafunctional macrocycle in Example 5 below is clearly shown in the index following the name. Here the ligand locants are necessary.

Note 10i. Any chelating ligand capable of binding with more than one set of donor atoms is described as *flexidentate*, cf., W. J. Stratton and D. H. Busch, *J. Am. Chem. Soc.*, **80**, 3191 (1958).

Example:

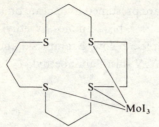

5. triiodo(1,4,8,12-tetrathiacyclopentadecane-$\kappa^3S^{1,4,8}$)molybdenum (Note 10j)

Well-established modes of chelation of the (1,2-ethanediyldinitrilo)tetraacetato-ligand are illustrated in Examples 6–9. These include didentate, tetradentate, and pentadentate examples. Note that a number of these Examples illustrate the names of complex anions, and that a complete compound name would require the specification of a cation.

Examples:

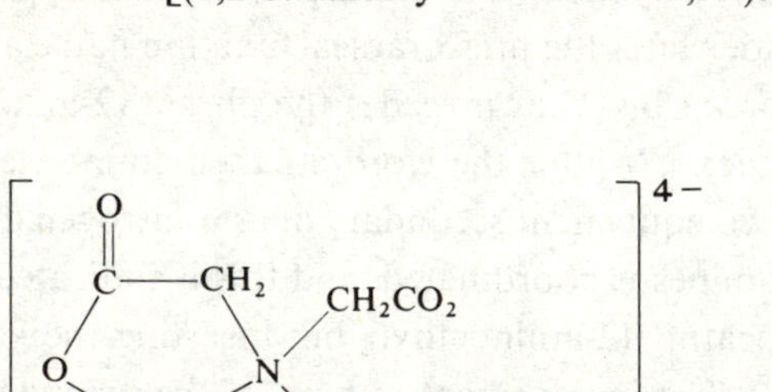

$$\left[\begin{array}{c} CH_2 - CH_2 \\ (O_2CCH_2)_2N \quad\quad N(CH_2CO_2)_2 \\ Pt^{II} \\ Cl \quad\quad Cl \end{array} \right]^{4-}$$

6. dichloro[(1,2-ethanediyldinitrilo-κ^2N,N')tetraacetato]platinate(4−)

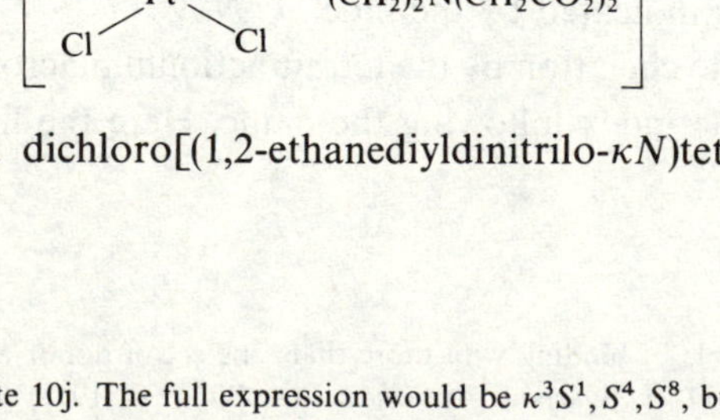

$$\left[\begin{array}{c} O \\ \| \\ C - CH_2 \quad CH_2CO_2 \\ O \quad\quad N \\ Pt^{II} \quad (CH_2)_2N(CH_2CO_2)_2 \\ Cl \quad Cl \end{array} \right]^{4-}$$

7. dichloro[(1,2-ethanediyldinitrilo-κN)tetraacetato-κO]platinate(II)

Note 10j. The full expression would be κ^3S^1,S^4,S^8, but the abbreviated form $\kappa^3S^{1,4,8}$ may be used.

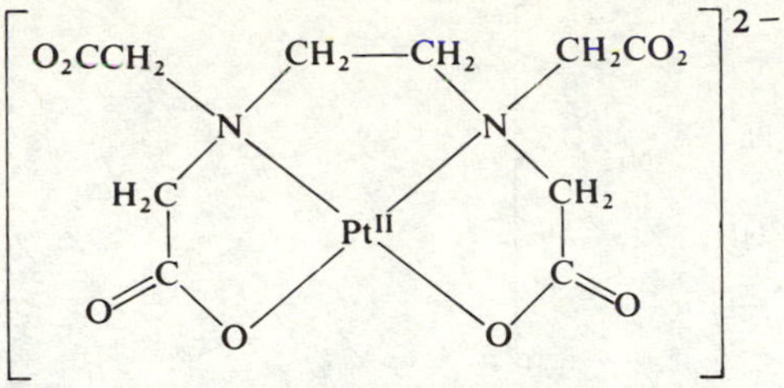

8. $[(1,2\text{-ethanediyldinitrilo-}\kappa^2N,N')\text{tetraacetato-}\kappa^2O,O'']\text{platinate}(2-)$

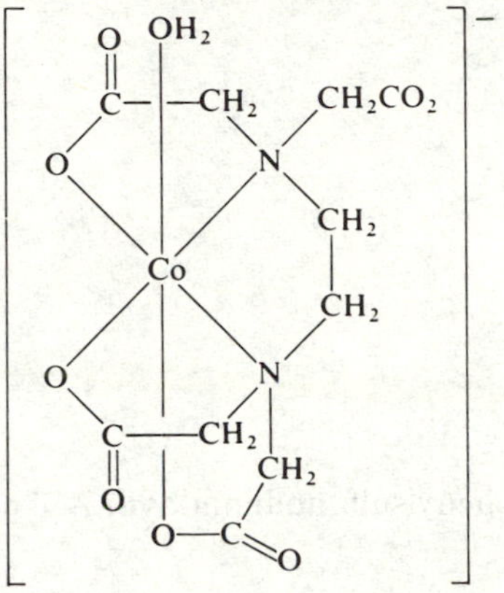

9. $\text{aqua}[(1,2\text{-ethanediyldinitrilo-}\kappa^2N,N')(\text{tetraacetato-}\kappa^3O,O'',O'''')]\text{cobaltate}(1-)$

The unknown flexidentate isomer (not presented above) in which one amino group is not coordinated while all four acetates are bound to a single metal ion would bear the ligand name (within the name of the complex).

$(1,2\text{-ethanediyldinitrilo-}\kappa N)\text{tetraacetato-}\kappa^4O,O'',O'''',O''''''$

The mixed sulfur–oxygen cyclic polyether, 1,7,13-trioxa-4,10,16-trithiacyclo-octadecane, might chelate to alkali metals only through its oxygen atoms and to second row transition atoms only through its sulfur atoms. The corresponding kappa indices are κ^3O^1,O^7,O^{13} and $\kappa^3S^4,S^{10},S^{16}$, respectively.

Examples 10, 11, and 12 below illustrate three modes of chelation of the ligand [N-[N-(2-aminoethyl)-N',S-diphenylsulfonodiimidoyl]benzamidine]. The abundance of heteroatoms in this structure makes the use of a special index to indicate ligating atoms especially helpful.

Examples:

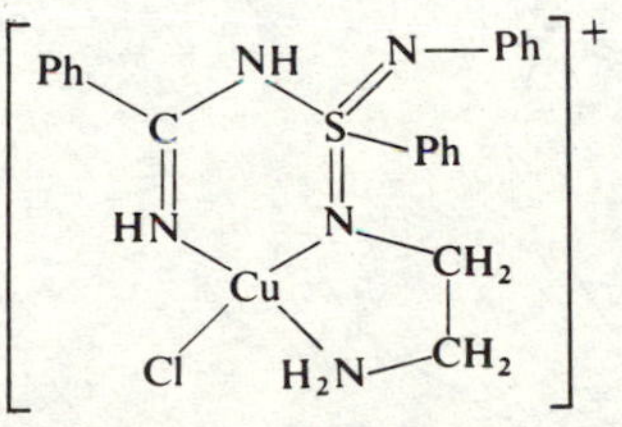

10. {N-[N-(2-amino-κN-ethyl)-N',S-diphenylsulfonodiimidoyl-κN]benzamidine-$\kappa N'$}chlorocopper(II) ion.

177

11. {N-[N-(2-amino-κN-ethyl)-N',S-diphenylsulfonodiimidoyl-$\kappa^2 N$,N']benzamidine}chlorocopper(II) ion.

12. {N-[N-(2-amino-κN-ethyl)-N',S-diphenylsulfonodiimidoyl-κN]benzamidine-κN}chlorocopper(II) ion.

The distinction between the names 10 and 12 rests on the conventional priming of the second nitrogen atom in the benzamidine functional group. The prime identifies the benzamidine nitrogen atom more remote from the 2-aminoethyl group, i.e., the imino-nitrogen atom.

The use of donor atom locants on the atomic symbols to indicate point-of-ligation is illustrated by the two isomeric didentate modes of binding of the macrocycle 1,4,7-triazacyclodecane (Examples 13 and 14). Conveying the formation of the five-membered chelate ring requires the index $\kappa^2 N^1$,N^4, while the six-membered chelate ring requires the index $\kappa^2 N^1$,N^7.

Examples:

13.   14.

Very complex structures can be treated as illustrated below.

15.

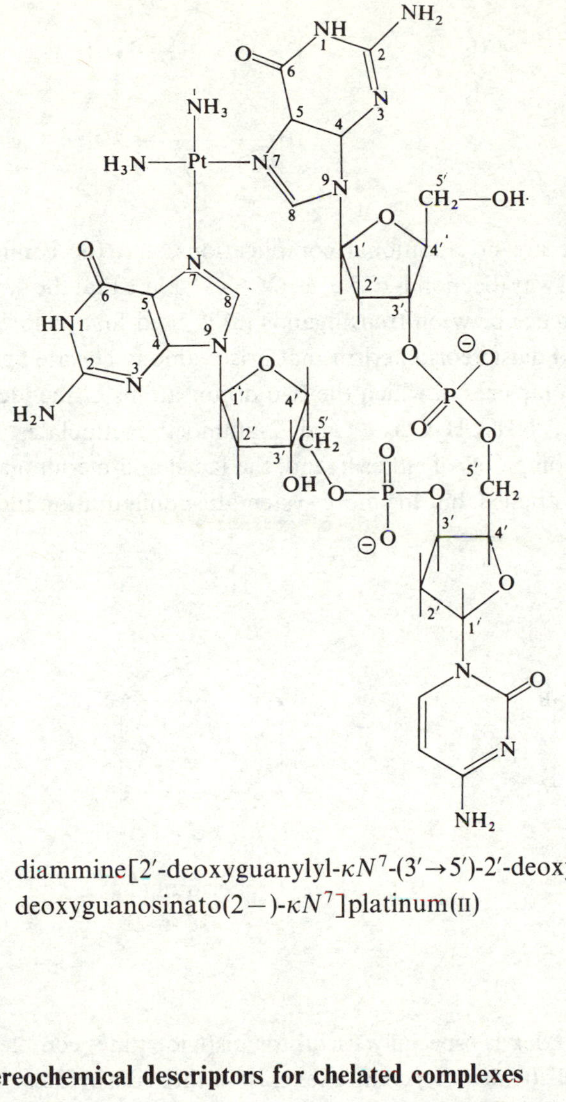

diammine[2'-deoxyguanylyl-κN^7-(3'→5')-2'-deoxycytidylyl(3'→5')-2'-
deoxyguanosinato(2−)-κN^7]platinum(II)

I-10.6.3   Stereochemical descriptors for chelated complexes

I-10.6.3.1   *General*

Stereochemical descriptors can be provided for compounds containing chelated ligands
but they involve considerations beyond those described above (Section I-10.5.1) for
complexes of monodentate ligands. The polyhedral symbol is determined as in the case of
monodentate ligand derivatives (Section I-10.5.2). Also, the priority numbers are assigned
to ligating atoms as for monodentate ligands (Section I-10.5.3.1). However, a general
treatment for the assignment of the configuration index requires the use of priming
conventions in order to provide a completely systematic treatment. Thus, for a particular
diastereoisomer of $[Co(NH_3)_2(NH_2CH_2CH_2NH_2)Br_2]^+$, the polyhedral symbol is OC-6
and the ligating atom priority numbers are as shown below.

Example:

1.

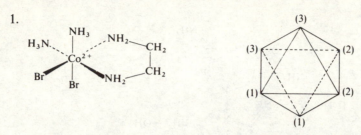

For this case, there are no additional complications, and the configuration index is assigned in the usual way (Section I-10.5.3) as *OC*-6-32. Note that the symbol representing the maximum difference between *trans* ligands (3) is cited first (before 2).

The classic case of diastereoisomerism that arises among chelate ligand derivatives is the tris(didentate) complexes in which the two donor atoms of the identical ligands are different. Glycinate, $NH_2CH_2CO_2^-$, and 2-aminoethanethiolate, $NH_2CH_2CH_2S^-$, illustrate this. For complexes of either ligand, the facial and meridional labels described previously could be applied, but the more systematic configuration indexes are *OC*-6-22 and *OC*-6-21.

Example:

2.

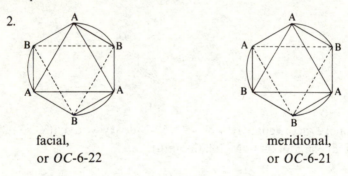

facial,   meridional,
or *OC*-6-22   or *OC*-6-21

I-10.6.3.2   *Priming convention*

The configuration index is especially useful for bis(tridentate) complexes and for more complicated cases. Bis(tridentate) complexes exist in three diastereoisomeric forms which serve to illustrate the utility of a priming convention. These isomers are represented below, along with their site symmetry symbols and configuration indexes. For Examples 1, 2, and 3, the two ligands are identical and the ligating-atom priority numbers are indicated.

Examples:

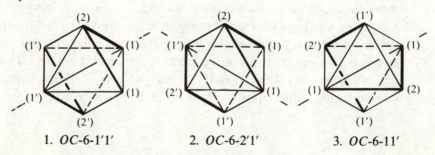

1. *OC*-6-1'1'   2. *OC*-6-2'1'   3. *OC*-6-11'

The priority numbers on one of the ligands are arbitrarily primed. In the procedure that we now describe, the primed number is assigned lower priority than the corresponding unprimed number, but a higher priority than the next higher unprimed number. Thus 1′ has lower priority than 1, but higher than 2. Note the clear distinction that is made between the isomers *OC*-6-1′1′ and *OC*-6-11′. This treatment can be applied to species such as *N*-(2-aminoethyl)-1,2-ethanediamine, commonly called diethylenetriamine, and to iminodiacetate. The technique also distinguishes between diastereoisomers for complexes of higher polydentate ligands as indicated in Examples 4, 5, and 6 for linear tetradentate ligands. In these cases, the donor atoms in half of the tetradentate ligand have been primed; examples of such ligands include *N*,*N*′-bis(2-aminoethyl)-1,2-ethanediamine, commonly called triethylenetetramine, and 2,2′-[1,2-ethanediylbis(nitrilomethylidyne)]diphenol.

Examples:

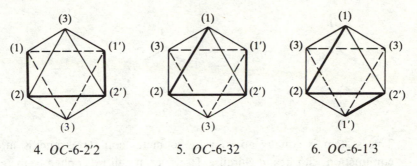

4. *OC*-6-2′2   5. *OC*-6-32   6. *OC*-6-1′3

Clearly, pentadentate and hexadentate ligands can be treated similarly, as shown in Examples 7 to 10. Examples 7 and 8 apply to diastereoisomers of classic linear hexadentate ligands, whereas Examples 9 and 10 apply to the branched structure of [[2-[(carboxymethyl)(2-hydroxyethyl)amino]ethyl]imino]diacetic acid.

Examples:

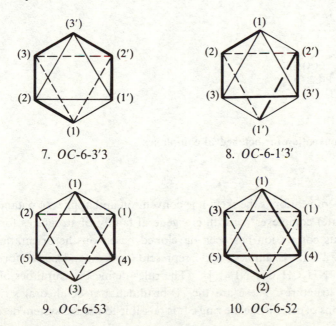

7. *OC*-6-3′3   8. *OC*-6-1′3′

9. *OC*-6-53   10. *OC*-6-52

I-10.7   CHIRALITY SYMBOLS

I-10.7.1   **Symbols *R* and *S***

There are two established and well-used systems for chirality symbols and these differ in fundamental ways. The first, the convention for chiral carbon atoms (and tetrahedral heteroatom centres) is dependent on the chemical constitution of the compound (see the *Nomenclature of Organic Chemistry*, 1979 edition, Section E-4) and uses the priority sequence cited in Section I-10.5.3.1. The chirality *R* is assigned if the cyclic sequence of priority numbers, proceeding from highest priority, is clockwise when the viewer is looking down the vector from the carbon atom to the substituent of lowest priority, as in Example 1. The converse applies for *S* (Example 2).

Examples:

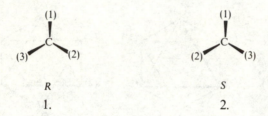

This system is equally appropriate to metal complexes and is most often used in conjunction with ligand chirality. However, it can be applied to metal centres and has been useful for pseudotetrahedral organometallic complexes when, for example, cyclopentadienyl ligands are treated as if they were monodentate ligands of high priority.

Example:

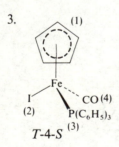

I-10.7.2   **Skew-line convention for octahedral complexes**

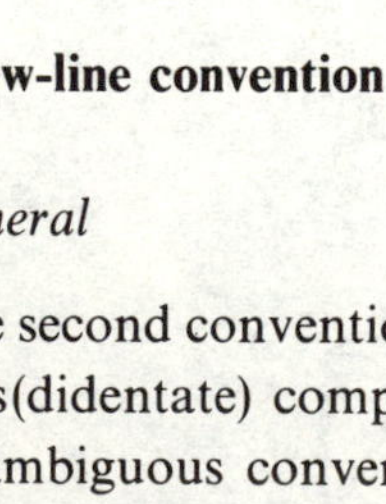

I-10.7.2.1   *General*

The second convention is the skew-line convention and applies to octahedral complexes. Tris(didentate) complexes constitute a general family of structures for which a useful unambiguous convention has been developed based on the orientation of skew lines which define a helix. Examples 1 and 2 represent the *delta* and *lambda* forms of a complex, such as $[Co(NH_2CH_2CH_2NH_2)_3]^{3+}$. The rules define the chiralities of two additional families of structures. These are the *cis*-bis(didentate) octahedral structures and the conformations of certain chelate rings. It is possible to use the system described below for complexes of higher polydentate ligands, but additional rules are required [M. Brorson, T. Damhus, and C. E. Schaeffer, *Inorg. Chem.*, **22**, 1569 (1983)].

Examples:

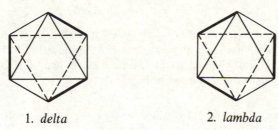

1. *delta*   2. *lambda*

I-10.7.2.2   *Basic principle of the convention*

Two skew-lines which are not orthogonal possess the property of having one, and only one, normal in common. They define a helical system, as illustrated in Figures I-10.1 and I-10.2. In Figure I-10.1, one of the skew-lines AA determines the axis of a helix upon a cylinder whose radius is equal to the length of the common normal NN to the two skew-lines AA and BB. The other of the skew-lines, BB, is a tangent to the helix at N and determines the pitch of the helix. In Figure I-10.2, the two skew-lines AA and BB are seen in projection onto a plane orthogonal to their common normal.

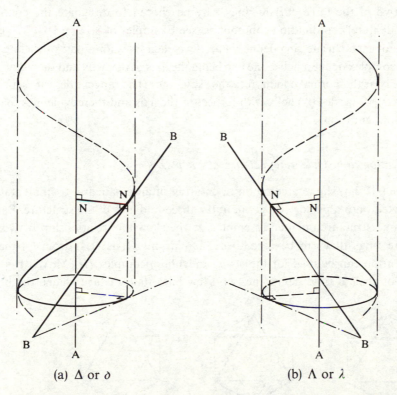

(a) Δ or δ   (b) Λ or λ

Figure I-10.1. Two skew lines AA and BB which are not orthogonal define a helical system. In the figure, AA is taken as the axis of a cylinder whose radius is determined by the common normal NN of the two skew-lines. The line BB is a tangent to the above cylinder at its crossing point with NN and defines a helix upon this cylinder. (a) and (b) illustrate a right- and left-handed helix respectively.

(a) Δ or δ   (b) Λ or λ

Figure I-10.2. The figure shows pairs of non-orthogonal skew-lines in projection upon a plane parallel to both lines. The full line BB is above the plane of the paper, the dotted line AA is below this plane. (a) corresponds to (a) of Figure I-10.1 and defines a right-handed helix. (b) corresponds to (b) of Figure I-10.1 and defines a left-handed helix.

Parts (a) of Figures I-10.1 and I-10.2 illustrate a right-handed helix to be associated with the Greek letter delta (Δ referring to configuration, δ to conformation). Parts (b) of Figures I-10.1 and I-10.2 illustrate a left-handed helix to be associated with the Greek letter lambda (Λ for configuration, λ for conformation). In view of the symmetry of the representation constituted by two skew-lines, the helix which the first line, say BB, determines around the second one, AA, has the same chirality as that which AA determines around BB. As one of the lines is rotated about NN with respect to the other, inversion occurs when the lines are parallel or perpendicular (Figure I-10.1).

I-10.7.2.3   *Application to tris(didentate) octahedral complexes*

Any two of the three chelate rings may be chosen to designate the configuration of tris(didentate) coordination compounds (see Examples of Section I-10.7.2.1). The donor atoms of each chelate ring define a line. Two such lines for a pair of chelate rings in the same complex define a helix, one line being the axis of the helix and the other a tangent of the helix at the normal common to the skew-lines. The tangent describes a right-handed (Δ) or a left-handed (Λ) helix with respect to the axis and thereby defines the chirality of that configuration.

I-10.7.2.4   *Application to bis(didentate) octahedral complexes*

Figure I-10.3(a) shows a common orientation of an octahedral tris(didentate) structure projected onto a plane orthogonal to the three-fold axis of the structure. Part (b) shows the same structure oriented to emphasize the skew-line relationship between a pair of chelate rings that can be used to define chirality. It is clear from (c) that the same convention can be used for the *cis*-bis(didentate) complex *cis*-$[M(AA)_2b_2]^{n+}$. The two pairs of chelate rings define the two skew-lines that, in turn, define the helix and the

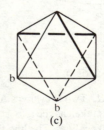

(a)   (b)   (c)

Figure I-10.3. Two orientations of a tris(didentate) structure, (a) and (b), to show the chiral relationship between these two species and bis(didentate) structure (c).

chirality of the substance. The procedure is precisely the same as described for the tris(didentate) case, but only a single pair of chelate rings is available.

I-10.7.2.5   *Application to conformations of chelate rings*

In order to assign the chirality of a ring conformation, the line AA in Figure I-10.2 is defined as that line joining the two ligating atoms of the chelate ring. The other line BB is that joining the two ring atoms which are neighbours to each of the ligating atoms. These two skew-lines define a helix in the usual way. The tangent describes a right-handed (δ) or a left-handed (λ) helix with respect to the axis and thereby defines the conformation in terms of the convention given in Figure I-10.1. Non-helical situations may also give rise to chirality when the chemical identities of the atoms are considered. For example, the chair and boat conformations of six-membered chelate rings are not chiral. However, if the two donor atoms are not identical then the chelate ring is chiral. Clearly the principles expounded in Section I-10.7.2.1 do not apply to such cases. The relationship between the convention of Figure I-10.2 and the usual representation of chelate ring conformation may be seen by comparing Figures I-10.2 and I-10.4.

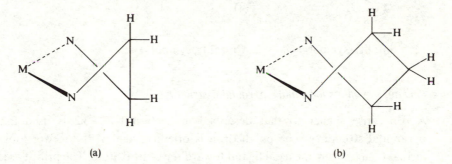

(a)   (b)

Figure I-10.4. δ-Conformation chelate rings: (a) five-membered; (b) six-membered.

I-10.7.3   **Chirality symbols based on the priority sequence**

I-10.7.3.1   *General*

The priority sequence used for the description of the chiralities of carbon atom centres is discussed in Section I-10.7.1. The same principles are readily extendable to geometries other than the tetrahedral [M. F. Brown, B. R. Cook, and T. E. Sloan, *Inorg. Chem.*, 7, 1563 (1978)]. Clearly there is no need to alter the rules in treating tetrahedral metal complexes. However, in order to avoid confusion, and to emphasize the unique aspects of the priority sequence system as applied to coordination polyhedra, the symbols *R* and *S* are replaced by the symbols *C* and *A* when applied to other polyhedra. It is also clear that no alternative notation is needed for chelate complexes where the skew-line convention is completely unambiguous.

The procedure for arriving at ligating atom priorities has been detailed in Section I-10.5.3.1.

I-10.7.3.2   *Chirality symbols for trigonal bipyramidal structures*

The procedure is applied as for tetrahedra, but modified because of the presence of a unique principal axis. The structure is oriented so that the viewer looks down the principal axis, with the ligand having the higher priority closer to the viewer and, accordingly, the axial ligand with the lower priority lies beyond the central atom. Using this orientation, the priority sequence of the three ligating atoms in the trigonal plane is examined. If the sequence proceeds from the highest priority to lowest priority in a clockwise fashion, the chirality symbol C is assigned. Conversely, if the sequence from highest to lowest priority (from lowest numerical index to highest numerical index) is anticlockwise, the symbol A is assigned.

Examples:

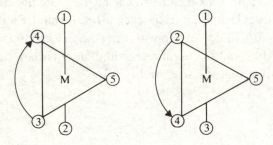

1. Chirality symbol $= C$ 2. Chirality symbol $= A$

I-10.7.3.3   *Chirality symbols for square pyramidal structures*

A procedure similar to that described in Section I-10.7.3.2 is used for tetragonal pyramidal structures. The polyhedron is oriented so that the viewer looks along the formal C_4 axis, from the axial ligand toward the central atom. The priority sequence then produces a C chirality for a clockwise decrease in priority (increase in the numbers themselves).

Examples:

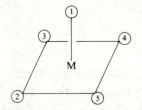 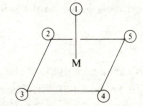

1. Chirality symbol $= C$ 2. Chirality symbol $= A$

I-10.7.3.4   *Chirality symbols for octahedral complexes*

The priority sequence based system is most useful for cases not handled easily by the skew-line conventions. However, configuration indexes (Section I-10.5.3) provide a basis for generating a priority sequence for any octahedral complex, and the A/C system can be applied generally. The chirality symbol is derived as for the square pyramid, except that

186

the principal axis is that axis containing the ligating atom of CIP priority 1, and that the atoms in the orthogonal coordination plane are viewed from the ligand having that highest priority (CIP priority 1). The chirality symbols C and A are then derived as for the tetragonal system (Section I-10.7.3.3).

Examples:

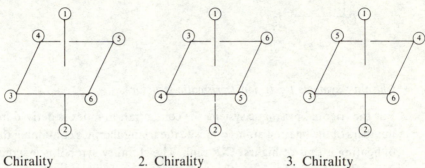

1. Chirality	2. Chirality	3. Chirality
symbol = C	symbol = A	symbol = C

Example 4 shows the compound $[CoBr_2(NH_3)_2(en)]^+$ which has the polyhedral symbol OC-6 and the configuration index 32. The chirality symbol is C.

Example:

4.

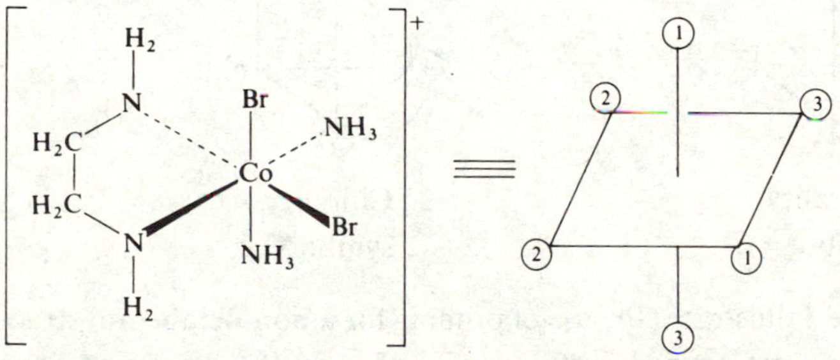

Example 5 shows the complex $[RuCl(CO)H(PMe_2Ph)_3]^{2+}$ which has the stereochemical descriptor OC-6-24-A. The chloro-ligand has priority 1.

Example:

5.

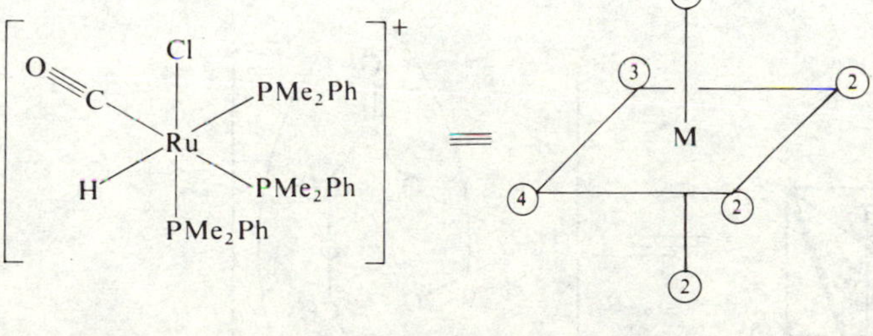

The assignment for polydentate ligands is illustrated by Example 6. This uses the priming convention developed in Section I-10.6.3.2.

Example:

6.

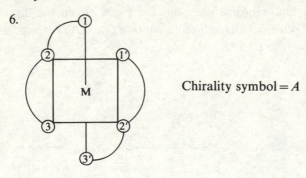

Chirality symbol $= A$

I-10.7.3.5   *Chirality symbols for trigonal prismatic structures*

For the trigonal prismatic system, the configuration index is derived from the CIP rank numbers of the ligating atoms opposite the triangular face containing the greater number of ligating atoms of highest CIP rank. The chirality symbol is assigned by viewing the trigonal prism from above the preferred triangular face and noting the direction of progression of the priority sequence for the less-preferred triangular face.

Examples:

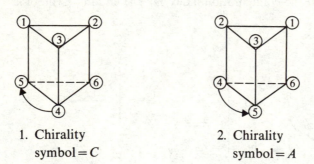

1. Chirality
   symbol $= C$

2. Chirality
   symbol $= A$

Example 3 illustrates the use of priming for a non-octahedral structure. The chirality designation is determined by the system of assigning primes to ligands. This uses the equivalent of the *trans maximum difference principle* (Section I-10.5.3.3). Specifically, the symbol 1 on the top face is placed above the symbol 1″ on the bottom face. This produces the sequence shown and the chirality symbol *C*. The stereochemical descriptor is *TPR-6-1″11′-C*.

Example:

3.

I-10.7.3.6 *Chirality symbols for other bipyramidal structures*

The procedure used for the trigonal bipyramid is appropriate for other bipyramidal structures, but an added complication is the need to identify the lexicographically lowest sequence in the orthogonal plane. This is the one having the lower number at the first point of difference when the numbers are compared digit by digit from one end to the other (see Section I-10.5.4.4). Example 1 has the stereochemical descriptor *PBPY*-7-12-11'1'33-*A*.

Example:

I-10.8 POLYNUCLEAR COMPLEXES

I-10.8.1 General

Polynuclear inorganic compounds exist in a bewildering array of structural types, such as ionic solids, molecular polymers, extended assemblies of oxoanions both of metals and non-metals, non-metal chains and rings, bridged metal complexes, and homo- and hetero-nuclear clusters. This section treats primarily the nomenclature of bridged metal complexes and homo- and hetero-nuclear clusters. Polynuclear complexes may have structures so large and extended as to make a rational structure-based nomenclature impractical. Their structures may be undefined or not suitably elucidated. Compositional nomenclature (Section I-1.3.3.2) is most suited to these circumstances.

I-10.8.2 Compositional nomenclature

In compositional nomenclature, names provide little structural information and their principal function is to convey the stoichiometric proportions of the various moieties present. Ligands are cited in the usual alphabetical order with appropriate numerical prefixes. Bridging ligands, as far as they can be specified, are indicated by the Greek letter μ appearing before the ligand name and separated by a hyphen. The whole term, e.g., μ-chloro, is separated from the rest of the name by hyphens, as in ammine-μ-chloro-chloro, etc., or by parentheses if more complex ligands are involved. If the bridging ligand occurs more than once and multiplicative prefixes are employed, the presentation is modified as in tri-μ-chloro-chloro, etc., or as in bis(μ-diphenylphosphido), etc., if more complex ligands are involved. The bridging index, the number of coordination centres connected by a bridging ligand, is indicated by a right subscript, μ_n, where $n \geq 2$. The bridging index 2 is not normally indicated. Bridging ligands are listed in alphabetical order along with the other ligands, but a bridging ligand is cited before a corresponding non-bridging ligand, as with di-μ-chloro-tetrachloro . . . . .Multiple bridging is listed in descending order of complexity, as shown by μ_3-oxo-di-μ-oxo-trioxo. . . . . For ligand names requiring enclosing marks, μ is contained within those enclosing marks (Note 10k). Central atoms are listed in alphabetical order after the ligands (Example 2). The number of central atoms of a given kind, if greater than one, is indicated by a numerical prefix (Example 1). For such anionic species, the suffix -ate and the number indicating the charge on the ion are added after the central atom list; the list of central atoms is then enclosed in parentheses (Example 3).

Examples:

1. $[Rh_3H_3\{P(OCH_3)_3\}_6]$ trihydridohexakis(trimethyl phosphite)trirhodium

2. $[CoCu_2Sn(CH_3)]\{\mu\text{-}(C_2H_3O_2)\}_2(C_5H_5)]$
            bis(μ-acetato)(cyclopentadienyl)(methyl)cobalt-dicoppertin

3. $[Fe_2Mo_2S_4(C_6H_5S)_4]^{2-}$ tetrakis(benzenethiolato)tetra-thio(diirondimolybdenum)ate(2−)

I-10.8.3   Structural nomenclature

I-10.8.3.1   *Dinuclear entities*

Bridging ligands are indicated in the same way as in compositional nomenclature above, unless the symmetry of the entity permits simpler names to be formed by modifying the symmetrical portions of the name with multiplicative prefixes.

Examples:
1. $[\{Cr(NH_3)_5\}_2(\mu\text{-OH})]Cl_5$ μ-hydroxo-bis(pentaamminechromium)(5+) pentachloride

2. $[[PtCl\{P(C_6H_5)_3\}]_2(\mu\text{-Cl})_2]$ di-μ-chloro-bis[chloro(triphenylphosphine)platinum]

3. $[\{Fe(NO)_2\}_2\{\mu\text{-}P(C_6H_5)_2\}_2]$ bis(μ-diphenylphosphido)bis(dinitrosyliron)

Metal–metal bonding may be indicated in names by italicized atomic symbols of the appropriate metal atoms, separated by a long dash and enclosed in parentheses, placed after the list of central atoms and before the ionic charge.

Examples:
4. $[Br_4ReReBr_4]^{2-}$ bis(tetrabromorhenate)(*Re—Re*)(2−)

5. $[Mn_2(CO)_{10}]$ bis(pentacarbonylmanganese)(*Mn—Mn*), or decacarbonyldimanganese(*Mn—Mn*)

Note 10k. The position of the bridging indicator μ and its placement within the enclosing marks for the ligand differ from that given in Section I-10.8.4 for single-strand coordination polymers.

I-10.8.3.2   *Unsymmetrical dinuclear entities*

Dinuclear coordination entities may be unsymmetrical, either because different types of metal atom are present or because of different patterns of ligation on similar metal atoms, or for both reasons. Heterodinuclear entities are numbered based on the priorities of the central elements listed in Table IV, the higher priority central atom (priority increases from right to left in Table IV) being numbered 1 even though they are cited in alphabetical order in the name. Unsymmetrical homo-dinuclear entities are numbered as follows: that central atom which gives the lowest locant set for ligands at the first point of difference takes priority (for these simple cases, this corresponds to the central atom with the larger number of ligands); if that fails, the alphabetical order of ligands establishes priority. The central atom with the greater number of ligands with initial letters earlier in the alphabet is numbered 1.

Where necessary, the symbol kappa, κ, with the italicized atomic symbol(s) of the donor(s) is employed to indicate the ligating atom(s) and their distribution. Bridging and unsymmetrical distribution of ligands is shown by the numerical locant of the central atom to which the ligand is bonded. The numerical locant of the central atom is placed on the line before the κ. Thus, (benzenethiolato-1κS) indicates that the sulfur atom of benzenethiolate is bonded to central atom number 1. A right superscript numeral is employed to denote the number of equivalent ligating atoms bonded to the specified central atom.

Example:

$$\overset{1}{}\overset{2}{}$$

1. $[(CO)_5Re-Co(CO)_4]$ nonacarbonyl-1κ^5C, 2κ^4C-cobaltrhenium(*Co—Re*)

Bridging is indicated by the μ prefix and where bridging is accomplished by different atoms of the same group, the ligating locants and symbols are separated by a colon, e.g., -μ-nitrito-1κN: 2κO-. In general, the hierarchy of marks is   , < : < ; but in this context we use the colon (:) only to indicate bridging. Thus, in Example 6 below, both the comma and the semicolon appear since no bridging is involved where the hierarchy of marks must be applied.

Examples:

$$\overset{1}{}\overset{2}{}$$

2. $[[IrCl_2(CO)\{P(C_6H_5)_3\}_2](HgCl)]$

      carbonyl-1κC-trichloro-1κ^2Cl,2κCl-
      bis(trisphenylphosphine-
      1κP)iridiummercury(*Hg—Ir*)

3. $[Cr(NH_3)_5(\mu\text{-}OH)Cr(NH_3)_4(NH_2CH_3)]Cl_5$

      nonaammine-μ-hydroxo-
      (methanamine)dichromium(5+)
      pentachloride

4. $[\{Co(NH_3)_3\}_2(\mu\text{-}OH)_2(\mu\text{-}NO_2)]Br_3$

      di-μ-hydroxo-μ-nitrito-κN: κO-
      bis(triamminecobalt)(3+) tribromide

5. [{Co(NH$_3$)$_3$}(μ-OH)$_2$(μ-NO$_2$){Co(C$_5$H$_5$N)(NH$_3$)$_2$}]Br$_3$

(1, 2 above the formula)

pentaammine-1$\kappa^3 N$,2$\kappa^2 N$-di-μ-hydroxo-
μ-nitrito-1κN: 2κO-(pyridine-
2κN)dicobalt(3+) tribromide

6. [Cu(2,2'-bpy)(H$_2$O)(μ-OH)$_2$Cu(2,2'-bpy)(SO$_4$)]

(1, 2 above the formula)

aqua-1κO-bis(2,2'-bipyridine)-
1$\kappa^2 N^1$,$N^{1'}$; 2$\kappa^2 N^1$,$N^{1'}$-di-μ-
hydroxo-[sulfato(2−)-2κO]dicopper(II)

7. [{Cu(C$_5$H$_5$N)}$_2$(μ-C$_2$H$_3$O$_2$)$_4$]

tetrakis(μ-acetato-
κO: $\kappa O'$)bis[(pyridine)copper(II)]

8. [Ni(NH$_3$)$_4$Cl{μ-(C$_2$H$_3$OS)}Ni(NH$_3$)$_3$Cl$_2$]

(1, 2 above the formula)

heptaammine-1$\kappa^4 N$,2$\kappa^3 N$-trichloro-
1κCl,2$\kappa^2 Cl$-(μ-thioacetato-2κO: 1κS)dinickel

9. Isomer of 8 with thioacetate   heptaammine-1$\kappa^4 N$,2$\kappa^3 N$-trichloro-
bound in opposite way   1κCl,2$\kappa^2 Cl$-(μ-thioacetato-1κO: 2κS)dinickel

I-10.8.3.3   *Trinuclear and larger structures*

The structural nomenclature of more complex polynuclear entities is based on the description of the central or fundamental structural unit and a logical procedure for numbering the atoms. Only the metal atoms are considered for this purpose. For nonlinear clusters, descriptors such as *tetrahedro* and *dodecahedro* traditionally have been used to describe a central structural unit or CSU. However, synthetic chemistry has now advanced far beyond the range of the limited CSU set associated with this usage (see below). These descriptors should only be used for simple cases.

A more comprehensive CSU descriptor and a numbering system, the CEP system, has been developed specifically for fully triangulated polyboron polyhedra (deltahedra) by J. B. Casey, W. J. Evans, and W. H. Powell, *Inorg. Chem.*, **20**, 1333 (1981). The CEP descriptors are systematic alternatives to the traditional descriptors for fully triangulated polyhedra (deltahedral) and are listed in Table I-10.8.

Table I-10.8 Structural descriptors

Number of atoms in CSU	Descriptor	Point group	CEP descriptor
3	triangulo	D_{3h}	
4	quadro	D_{4h}	
4	tetrahedro	T_d	[T_d-(13)-Δ^4-*closo*]
5		D_{3h}	[D_{3h}-(131)-Δ^6-*closo*]
6	octahedro	O_h	[O_h-(141)-Δ^8-*closo*]
6	triprismo	D_{3h}	
8	antiprismo	S_6	
8	dodecahedro	D_{2d}	[D_{2d}-(2222)-Δ^6-*closo*]
8	hexahedro (cube)	O_h	
12	icosahedro	I_h	[I_h-(1551)-Δ^{20}-*closo*]

Numbering of the CSU is based on locating a reference axis and planes of atoms perpendicular to the reference axis. The reference axis is the axis of highest rotational symmetry. Select that end of the reference axis with a single atom (or smallest number of atoms) in the first plane to be numbered. Orient the CSU so that the first position to receive a locant in the first plane with more than one atom is in the twelve o'clock position. Assign locant numbers to the axial position or to each position in the first plane, beginning at the twelve o'clock position and moving in either the clockwise or anticlockwise direction. From the first plane move to the next position and continue numbering in the same direction (clockwise or anticlockwise), always returning to the twelve o'clock position or the position nearest to it in the forward direction before assigning locants in that plane. Continue numbering in this manner until all positions are numbered. A complete discussion of numbering deltahedra is given in the reference cited above. The complete descriptor for the CSU should appear just before the central atom list. Where structurally significant, metal–metal bonds may be indicated (Section I-10.8.3.1).

I-10.8.3.4   *Symmetrical central structural units*

Central structural units may be identified specifically and numbered for nomenclature as described in Section I-10.8.3.3. However, many symmetrical CSUs may not require a full set of locants in the name because compounds based upon them do not exhibit isomerism.

Locants for bridging ligands are cited as for dinuclear entities. Compound locants will at times be necessary for monoatomic bridges in more complicated polynuclear entities. For these entities the locants are cited on line before the ligand indicator κ and separated by a colon, e.g., tri-μ-chloro-$1:2\kappa^2Cl;1:3\kappa^2Cl;2:3\kappa^2Cl$-. Note that because of the use of the colon, sets of bridge locants are separated by semicolons. This practice applies the hierarchy of punctuation marks cited in Section I-10.8.3.2.

Examples:

1. $[\{Co(CO)_3\}_3(\mu_3\text{-}CI)]$

   nonacarbonyl-(μ_3-iodomethylidyne)-*triangulo*-tricobalt(3 *Co—Co*)

2. $Cs_3[Re_3Cl_{12}]$

   caesium dodecachloro-*triangulo*-trirhenate(3 *Re—Re*)(3−)

3. $[Cu_4I_4\{P(C_2H_5)_3\}_4]$

   tetra-μ_3-iodo-tetrakis(triethylphosphine-*tetrahedro*-tetracopper, or tetra-μ_3-iodo-tetrakis(triethylphosphine) [T_d-(13)-Δ^4-*closo*]tetracopper

4.

   penta-μ-carbonyl-$1:2\kappa^2C;1:4\kappa^2C; 2:3\kappa^2C; 2:4\kappa^2C; 3:4\kappa^2C$-heptacarbonyl-$1\kappa^3C,2\kappa C,3\kappa^2C,4\kappa C$-*tetrahedro*-tetracobalt(4 *Co—Co*), or penta-μ-carbonyl-$1:2\kappa^2C;1:4\kappa^2C;2:3\kappa^2C; 2:4\kappa^2C; 3:4\kappa^2C$-heptacarbonyl-$1\kappa^3C,2\kappa C,3\kappa^2C,4\kappa C$-[$T_d$-(13)-$\Delta^4$-*closo*]tetracobalt(4 *Co—Co*)

5. $[Mo_6S_8]^{2-}$

octa-μ_3-thio-*octahedro*-
hexamolybdate$(2-)$,
or octa-μ_3-thio-$[O_h$-(141)-Δ^8-
closo]hexamolybdate$(2-)$

6.

tetra-μ_3-iodo-
tetrakis[trimethylplatinum(IV)],
or tetra-μ_3-iodo-dodecamethyl-
$1\kappa^3C,2\kappa^3C,3\kappa^3C,4\kappa^3C$-*tetrahedro*-
tetraplatinum(IV), or tetra-μ_3-iodo-
dodecamethyl-$1\kappa^3C,2\kappa^3C,3\kappa^3C,4\kappa^3C$-
$[T_d$-(13)-Δ^4-*closo*]tetraplatinum(IV)

7. $[Be_4(\mu\text{-}C_2H_3O_2)_6(\mu_4\text{-}O)]$

hexakis(μ-acetato-κO:$\kappa O'$)-μ_4-oxo-
tetrahedro- tetraberyllium,
or hexakis(μ-acetato-κO:$\kappa O'$)-μ_4-oxo-
$[T_d$-(13)-Δ^4-*closo*]tetraberyllium

8. $[\{Hg(CH_3)\}_4(\mu_4\text{-}S)]^{2+}$

μ_4-thio-tetrakis(methylmercury)$(2+)$ ion,
or tetramethyl-$1\kappa C,2\kappa C,3\kappa C,4\kappa C$-$\mu_4$-
thio-*tetrahedro*-tetramercury$(2+)$ ion,
or tetramethyl-$1\kappa C,2\kappa C,3\kappa C,4\kappa C$-$\mu_4$-thio
$[T_d$-(13)-Δ^4-*closo*]tetramercury$(2+)$ ion

9.

octacarbonyl-$1\kappa^4C,2\kappa^4C$-bis(triphenylphosphine-$3\kappa P$)-
triangulo-diironplatinum$(Fe\text{---}Fe)(2\,Fe\text{---}Pt)$

I-10.8.3.5   *Unsymmetrical central structural units*

Central atoms in chain, branched-chain, and cyclic polynuclear structures are numbered consecutively from one end along the path containing the greatest number of central atoms. The appropriate end to start numbering is selected as follows: by the priorities of the central elements listed in Table IV, the highest priority being given to the element coming last in that sequence; if that is not sufficient, that central atom numbering which leads to the assignment of the lowest locant set for ligands, at the first point of difference, takes priority; and if that fails, the alphabetical order of ligands establishes priority. The favoured terminal central atom is numbered 1. When needed, the locant precedes the atom name in the central atom list, within the name (see Example 3). The ligand indicator kappa, κ, with central atom locant and italicized donor atom symbol is used where necessary to indicate the positions of the ligating atoms.

Examples:

1.   5   4   3   2   1
   SiH₃–Si(CH₃)₂–SiCl₂–SiH₂–SiClH₂

   trichloro-$1\kappa Cl,3\kappa^2 Cl$-
   heptahydridodimethyl-$4\kappa^2 C$-
   pentasilicon (Note (10l)

2. $[Os_3(SiCl_3)_2(CO)_{12}]$

   dodecacarbonyl-$1\kappa^4 C,2\kappa^4 C,3\kappa^4 C$-
   bis(trichlorosilyl)-$1\kappa Si,3\kappa Si$-
   triosmium(2 *Os—Os*)

3.

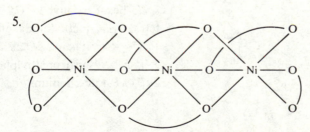

   tricarbonyl-$1\kappa C,2\kappa C,3\kappa C$-$\mu$-
   chloro-$1:2\kappa^2 Cl$-chloro-$3\kappa Cl$-
   bis{μ_3-bis[(diphenylphosphino)-$1\kappa P'$:
   $3\kappa P''$-methyl]phenylphosphine-
   $2\kappa P$}trirhodium(1+) chloride

4.

   hexaammine-$2\kappa^3 N,3\kappa^3 N$-aqua-
   $1\kappa O$-[μ_3-(1,2-ethanediyldinitrilo-
   $1\kappa^2 N,N'$)-tetraacetato-
   $1\kappa^3 O^1,O^2,O^3:2\kappa O^4:3\kappa O^{4'}$]-di-$\mu$-
   hydroxo-$2:3\kappa^4 O$-1-chromium-
   2,3-dicobalt(3+) triperchlorate

5.

bis(μ_3-2,4-pentanedionato-$1:2\kappa^2 O^2;2:3\kappa^2 O^4$)bis($\mu$-2,4-pentanedionato)-
$1\kappa O^2,1:2\kappa^2 O^4;3\kappa O^4,2:3\kappa O^2$-bis(2,4-pentanedionato)-$1\kappa^2 O^2,O^4;3\kappa^2 O^2,O^4$-trinickel

Central atoms in bridged cyclic structures are numbered as follows: by use of Table IV, the highest priority is assigned to the central atom given last in that sequence; second, if this is not sufficient, higher priority is given by the lowest locant set for the ligands bound to a given central atom at the first point where a difference of coordination shell arises; and if that fails, the alphabetical order of ligands establishes priority. The central

Note 10l. 1, 3, 3, 4, 4 is a lower locant set at the first point of difference as compared to 2, 2, 3, 3, 5.

atom with the greatest number of alphabetically preferred ligands has highest priority. The prefix *cyclo-* may be used for monocyclic compounds. The *cyclo-* prefix is italicized and cited before all ligands.

Examples:

6. $[Pt_3(NH_3)_6(\mu\text{-}OH)_3]^{3+}$ *cyclo-*tri-μ-hydroxo-tris(diammineplatinum)(3+), or hexaamminetri-μ-hydroxo-*triangulo-*triplatinum(3+)

7.

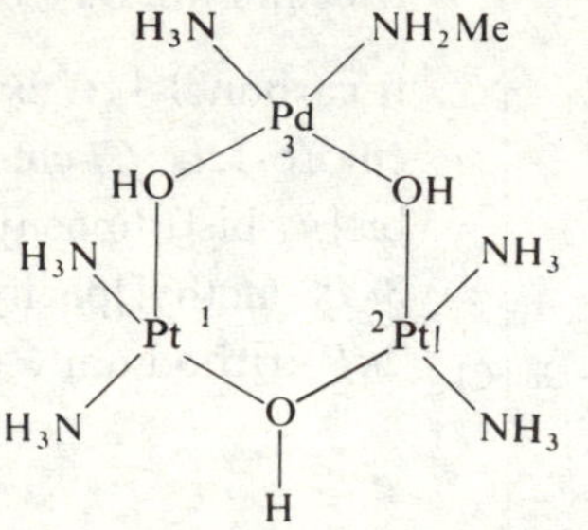

*cyclo-*pentaammine-$1\kappa^2 N,2\kappa^2 N,3\kappa N$-tri-$\mu$-hydroxo-$1{:}2\kappa^2 O;1{:}3\kappa^2 O;2{:}3\kappa^2 O$-(methylamine-$3\kappa N$)palladiumdiplatinum(3+)

8.

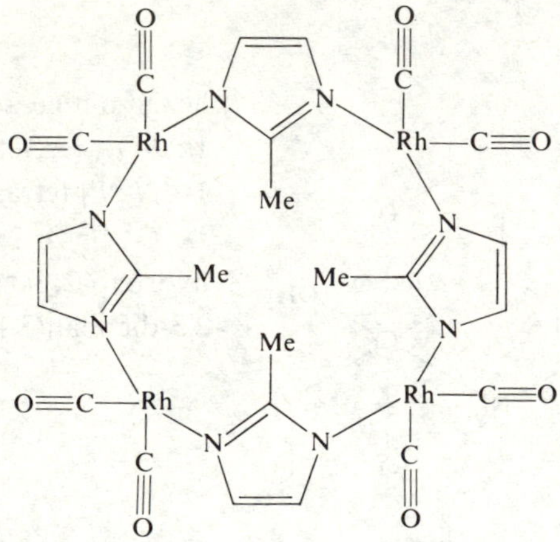

*cyclo-*tetrakis(μ-2-methylimidazolato-$\kappa N^1{:}\kappa N^3$)tetrakis(dicarbonylrhodium)

9.

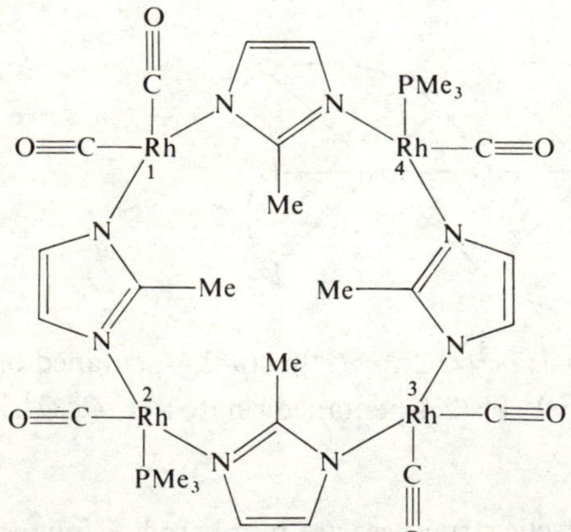

*cyclo-*hexacarbonyl-$1\kappa^2 C,2\kappa C,3\kappa^2 C,4\kappa C$-tetrakis($\mu$-2-methylimidazolato)-$1{:}2\kappa^2 N^1{:}N^3;1{:}4\kappa^2 N^1{:}N^3;2{:}3\kappa^2 N^1{:}N^3,$-$3{:}4\kappa^2 N^1{:}N^3$-bis(trimethylphosphine)-$2\kappa P,4\kappa P$-tetrarhodium

When the prefix *cyclo-* is used in the names of metal–metal bonded entities, the symbols indicating the presence of the metal–metal bonds are required (Example 10).

Example:

10. $[Os_3(CO)_{12}]$ *cyclo-*dodecacarbonyl-$1\kappa^4 C,2\kappa^4 C,3\kappa^4 C$-triosmium(3 *Os—Os*)

I-10.8.4   **Single-strand coordination polymers**

Where bridging causes an indefinite extension of the structure, compounds are named on the basis of the repeating unit. Thus the compound having the composition represented by the formula $ZnCl_2 \cdot NH_3$ has the structure given below and should be named as a polymer.

$$\cdots\cdots\overset{\overset{\textstyle NH_3}{\textstyle |}}{\underset{\underset{\textstyle Cl}{\textstyle |}}{Zn}}—Cl—\overset{\overset{\textstyle NH_3}{\textstyle |}}{\underset{\underset{\textstyle Cl}{\textstyle |}}{Zn}}—Cl—\overset{\overset{\textstyle NH_3}{\textstyle |}}{\underset{\underset{\textstyle Cl}{\textstyle |}}{Zn}}—Cl—\overset{\overset{\textstyle NH_3}{\textstyle |}}{\underset{\underset{\textstyle Cl}{\textstyle |}}{Zn}}—Cl\cdots\cdots$$

A doubly bridged polymeric structure is found for $PdCl_2$. Both these materials are single-strand polymers.

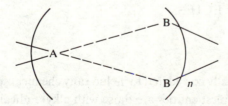

Regular single-strand polymers are characterized by a *constitutional repeating unit*, or CRU, that is joined at both ends through single atoms $-(A-B)_n-$. In $ZnCl_2 \cdot NH_3$, A and B are Zn and Cl, respectively. Nomenclature practices for these species are documented in the 1985 IUPAC Recommendations from the Commission on Macromolecular Nomenclature and the Commission on Nomenclature of Inorganic Chemistry (Note 10 m). Quasi-single-strand polymers are similarly characterized, but only one end is connected through a single atom to the next identical CRU. In $PdCl_2$, A and B are Pd and Cl, respectively. The nomenclature rules defining the CRU and representing the polymer in terms of the CRU are not totally consistent with the rules given above for polynuclear complexes. Only an elementary account is given here.

The first polymer cited above, $ZnCl_2 \cdot NH_3$, is named *catena*-poly[(amminechlorozinc)-μ-chloro]. Similarly, the name for palladium(II) chloride polymer is *catena*-poly[palladium(II)-μ-dichloro]. Plainly, these names are inverted with respect to the usual practice in inorganic chemistry.

Essential elements of the nomenclature for single-strand polymeric coordination compounds include the selection of the constitutional repeating unit (CRU), its orientation so that the highest priority central atom (Table IV) is listed first, the naming of that part of the CRU that does not include bridging ligands according to coordination nomenclature practices, and finally the naming of bridging ligands, separated by the bridging indicator μ (Note 10n).

Note 10m. *Pure Appl. Chem.*, **57**, 149 (1985).

Note 10n. The position of the bridging indicator μ in the names of coordination polymers and its placement outside the ligand-enclosing marks differs from the style recommended for polynuclear complexes in Section I-10.8.2.

Examples:

1.

catena-poly[[thiourea-*S*)silver]-μ-chloro]

2.

catena-poly[silver-μ-(cyano-*N:C*)] (Note 10o)

3.

catena-poly[(diamminedibromoplatinum)-μ-bromo]

4.

catena-poly[nickel-μ-[dithiooxamidato(2−)-κ*N*,κ*S*:κ*N'*,κ*S'*]]

5.

catena-poly[zinc-μ-[2,5-dihydroxy-*p*-benzoquinonato(2−)-*O*^{1},*O*^{2}:*O*^{4},*O*^{5}]]

I-10.9   ORGANOMETALLIC SPECIES

I-10.9.1   General

Organometallic entities are usually considered to include any chemical species containing a carbon–metal bond. The simplest entities are those with alkyl radical ligands, such as diethylzinc. In general, ligands bound by a single carbon atom to metals are named by the customary substituent group names, though these ligands must be treated as anions in order to calculate oxidation numbers. In any case, the designation is arbitrary. Ligands conventionally treated as having metal–donor double bonds (alkylidenes) and triple bonds (alkylidynes) are also given substituent group (radical) names.

Examples:
1. $[Hg(CH_3)_2]$ dimethylmercury
2. $MgBr[CH(CH_3)_2]$ bromo(isopropyl)magnesium
3. $[Tl(CN)(C_6H_5)_2]$ cyanodiphenylthallium
4. $[Fe(CH_3CO)I(CO)_2\{P(CH_3)_3\}_2]$

     acetyldicarbonyliodobis(trimethylphosphine)iron

Note 10o. This is not named *catena*-poly[silver-μ-(cyano-*C:N*)]. See Note 10m cited above on macromolecular nomenclature for details.

5. $[W\{CC_6H_5)(CH_3O)\}(CO)_4(NCCH_3)]$

                            (acetonitrile)tetracarbonyl(α-methoxybenzylidene)-
tungsten

6. $[W(SeC_6H_5)(CO)_4[C\{N(C_2H_5)_2\}]]$

                            (benzeneselenolato)tetracarbonyl-
[(diethylamino)methylidyne]tungsten

7. $[Pt\{\overline{C(O)CH(C_6H_5)CH(C_6H_5)}\}\{P(C_6H_5)_3\}_2]$

                            (1-oxo-2,3-diphenyltrimethylene)-$\kappa C^1,\kappa C^3$-
bis(triphenylphosphine)platinum

8. $[Co\{Si(CH_3)_3\}\{P(C_2H_5)_3\}(CO)_3]$

                            tricarbonyl(triethylphosphine)(trimethylsilyl)cobalt

Table I-10.9 contains a list of organic radical names for use in naming coordination compounds.

Table I-10.9 Organic radical names used in coordination nomenclature

Formula of radical	Systematic name	Alternative name[a]
CH_3-	methyl	
CH_3CH_2-	ethyl	
$CH_3CH_2CH_2-$	propyl	
$\begin{array}{c}CH_3\\ \diagup\\ CH-\\ \diagdown\\ CH_3\end{array}$	1-methylethyl	isopropyl
$CH_2{=}CHCH_2-$	2-propenyl	allyl
$CH_3CH_2CH_2CH_2-$	butyl	
$\begin{array}{c}CH_3\\ \diagup\\ CHCH_2-\\ \diagdown\\ CH_3\end{array}$	2-methylpropyl	isobutyl
$\begin{array}{c}CH_3\\ \vert\\ CH_3CH_2CH-\end{array}$	1-methylpropyl	*sec*-butyl
$\begin{array}{c}CH_3\\ \vert\\ CH_3C-\\ \vert\\ CH_3\end{array}$	1,1-dimethylethyl	*tert*-butyl
$\begin{array}{c}H_2C\\ \diagup\ \diagdown\\ \quad\ \ CH-\\ \diagdown\ \diagup\\ H_2C\end{array}$	cyclopropyl	
$\begin{array}{c}H_2C{-}CH_2\\ \vert\qquad\ \vert\\ H_2C{-}CH-\end{array}$	cyclobutyl	
$\begin{array}{c}H_2C{-}CH_2\\ \vert\qquad\quad CH-\\ H_2C{-}CH_2\end{array}$	cyclopentyl	

Table I-10.9 (*continued*)

Formula of radical	Systematic name	Alternative name[a]
C_5H_5-	cyclopentadienyl	
C_6H_5-	phenyl	
$C_6H_5CH_2-$	benzyl	
$C_{10}H_7-$	1- or 2-naphthalenyl	1- or 2-naphthyl
C_9H_7-	1H–indenyl	indenyl
$C_{10}H_{17}-$	1,7,7-trimethylbicyclo[2.2.1]hept-2-yl	2-bornyl
$CH_3\overset{\overset{O}{\|\|}}{C}-$	acetyl	
SiH_3	silyl	
SnH_3	stannyl	
GeH_3	germyl	
$CH_3\underset{\underset{CH_3}{\|}}{\overset{\overset{CH_3}{\|}}{C}}CH_2-$	2,2-dimethylpropyl	neopentyl
$CH_3CH_2\overset{\overset{O}{\|\|}}{C}-$	propionyl	
$CH_3CH_2CH_2\overset{\overset{O}{\|\|}}{C}-$	butyryl	
$CH_2=$	methylene	
$CH_3CH=$	ethylidene	
$CH_3CH_2CH=$	propylidene	
$CH_2=CHCH=$	2-propenylidene	allylidene
$\underset{CH_3}{\overset{CH_3}{>}}C=$	1-methylethylidene	
$\underset{H_2C}{\overset{H_2C}{>}}C=$ (cyclopropyl ring)	cyclopropylidene	
cyclobutyl ring $=$	cyclobutylidene	
2,4-cyclopentadien ring $=$	2,4-cyclopentadien-1-ylidene	
$C_6H_5CH=$	benzylidene	
$CH\equiv$	methylidyne	
$CH_3C\equiv$	ethylidyne	
$CH_3CH_2-C\equiv$	propylidyne	
$C_6H_5C\equiv$	benzylidyne	
$-CH_2CH_2-$	1,2-ethanediyl	ethylene
$-CH_2CH_2CH_2-$	1,3-propanediyl	trimethylene
$-CH_2CH_2CH_2CH_2-$	1,4-butanediyl	tetramethylene

Table I-10.9 (*continued*)

Formula of radical	Systematic name	Alternative name[a]
$CH_2=CH-$	ethenyl	vinyl
$\begin{array}{c} -CHCH_2- \\ \mid \\ CH_3 \end{array}$	1-methyl-1,2-ethanediyl	propylene
$CH\equiv C-$	ethynyl	

[a]The alternative names are normally preferred.

Ligands chelated to metals through a neutral heteroatom and a carbon atom are given the customary substituent (radical) names, and the heteroatom bonding is indicated by the italicized donor atom symbols or the κ notation.

Examples:

9. $[\overline{Pd\{C_6H_4CH_2N(CH_3)_2\}Cl_2}]^-$ dichloro{2-[(dimethylamino)methyl]phenyl-C^1,N}palladate(II), or dichloro{2-[(dimethylamino-κN)methyl]phenyl-κC^1}palladate(II)

10. $[\overline{Mn\{C_6H_4NN(C_6H_5)\}}(CO)_4]$ tetracarbonyl[2-(phenylazo)phenyl-C^1,N^2]manganese, or tetracarbonyl[2-(phenylazo-κN^2)phenyl-κC^1]manganese

Names of ligands bonded to metals by a carbon atom and an anionic heteroatom are given the -ato ending and the apparent anionic charge is indicated by the charge number. The bonding atoms are indicated by the donor-atom symbols or the κ notation.

Examples:

11. $[\overline{Ni\{CH_2CH_2C(O)O}\}\{(C_6H_5)_2PCH_2CH_2P(C_6H_5)_2\}]$ [1,2-ethanediylbis(diphenylphosphine-P)][propanoato(2−)-C^3,O]nickel(II), or [1,2-ethanediylbis(diphenylphosphine-κP)][propanoato(2−)$\kappa C^3,\kappa O$]nickel(II)

12. $[\overline{Ni(CH_2CH_2C(O)N}CH_3)\{(C_6H_5)_2PCH_2CH_2P(C_6H_5)_2\}]$ [1,2-ethanediylbis(diphenylphosphine-P)][N-methylpropanamidato(2−)-C^3,N]nickel, or [1,2-ethanediylbis(diphenylphosphine-κP)][N-methylpropanamidato(2−)-$\kappa C^3,\kappa N$]nickel

I-10.9.2 **Complexes with unsaturated molecules or groups**

Since the first reported synthesis of ferrocene, the numbers and variety of organometallic compounds with unsaturated organic ligands have increased enormously. Further complications arise because alkenes, alkynes, imides, diazenes, and other unsaturated ligand systems such as cyclopentadienyl, $C_5H_5^-$, 1,3-butadiene, C_4H_6, and cycloheptatrienylium, $C_7H_7^+$, may be formally anionic, neutral, or cationic. The structures and bonding in some instances may be complex or ill-defined. For these cases, names indicating stoichiometric composition, constructed in the usual manner, are convenient. The ligand names are arranged in alphabetical order, and followed by central atom names, also in alphabetical order. Bonding notation is not given.

Examples:

1. $[PtCl_2(C_2H_4)(NH_3)]$

   amminedichloro(ethene)platinum

2. $[Hg(C_5H_5)_2]$

   bis(cyclopentadienyl)mercury

3. $[Fe_4Cu_4(C_5H_5)_4\{[(CH_3)_2N]C_5H_4\}_4]$

   tetrakis(cyclopentadienyl)tetrakis[(dimethylamino)cyclopentadienyl]-tetracoppertetrairon

The unique nature of the bonding of hydrocarbon and other π-electron systems to metals and the complex structures of these entities have rendered conventional nomenclature impotent. To accommodate the problems posed by the bonding and structures, the hapto nomenclature symbol was devised [F. A. Cotton, *J. Am. Chem. Soc.*, **90**, 6230 (1968)]. The hapto symbol, η (Greek eta), with numerical superscript, provides a topological description by indicating the connectivity between the ligand and the central atom.

The symbol η is prefixed to the ligand name, or to that portion of the ligand name most appropriate, to indicate the connectivity, as in (η^2-ethenylcyclopentadiene) and (ethenyl-η^5-cyclopentadienyl). The right superscript numerical index indicates the number of ligating atoms *in the ligand* which bind to the metal (Examples 4 to 10). In the entities in which the superscript index is not sufficient to specify a unique structure, locants of the bonding ligating atoms are placed before η. When the numerical locants of the ligating atoms are used, the locants and η are enclosed in parentheses, as in (1,2,3-η)- (Examples 10, 11, and 13). When locants are used, the superscript to η is omitted since it is clearly superfluous. In polynuclear entities the numerical locant of the central atom is given before η, and η is always enclosed in parentheses, as in 1(η^5)- and 2(1,2,3-η)-. These principles are further exemplified in Examples 13 to 17.

Examples:

4. $[Fe(CO)_3(C_4H_6SO)]$ tricarbonyl(η^2-2,5-dihydrothiophene 1-oxide-κO)iron

5. $[Cr(C_3H_5)_3]$ tris(η^3-allyl)chromium

6. $[Cr(CO)_4(C_4H_6)]$ tetracarbonyl(η^4-2-methylene-1,3-propanediyl)chromium

7. $[PtCl_2(C_2H_4)(NH_3)]$ amminedichloro(η^2-ethene)platinum

8. $[Fe(CO)_3(C_7H_8)]$ (η^4-bicyclo[2.2.1]hepta-2,5-diene)tricarbonyliron

9. [U(C$_8$H$_8$)$_2$]   bis(η^8-1,3,5,7-cyclooctatetraene)uranium

10.   dicarbonyl(η^5-cyclopentadienyl)-[(1,2,3-η)-2,4,6-cycloheptatrienyl]molybdenum

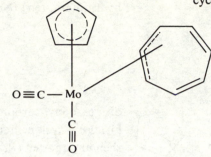

11.   dicarbonyl(η^5-cyclopentadienyl)[(4,5-η,κC^1)-2,4,6-cycloheptatrienyl]molybdenum] (Note 10p)

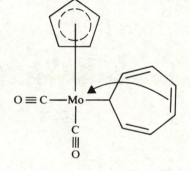

12.   tricarbonyl(η^7-cycloheptatrienylium)molybdenum(1+)

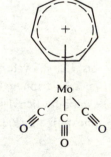

In Example 13, the colon is used as in other bridged structures (see Section I-10.8.3.2).

13.   [μ-(1,2,3,4-η: 5,6,7,8-η)]-1,3,5,7-cyclooctatetraene]bis(tricarbonyliron)

Note 10p. Others have used η^1 where κC is used in examples such as these. For single carbon–metal bonds, the use of κC is more appropriate.

14.

$[\mu\text{-}1\kappa C\text{:}2(\eta^5)\text{-cyclopentadienylidene}]\text{-}$
$[\mu\text{-}2\kappa C\text{:}1(\eta^5)\text{-cyclopentadienylidene}]\text{-}$
$\text{bis}[(\eta^5\text{-cyclopentadienyl})\text{-}$
hydridotungsten] (Note 10q)

15.

di-μ-carbonyl-carbonyl-$2\kappa C$-tris-
$[1,1,2(\eta^5)\text{-cyclopentadienyl}]\text{-}$
rheniumtungsten (Re—W)

16.

$[\mu\text{-}(1,2,3,3a,8a\text{-}\eta\text{:}4,5,6)\text{-azulene}]\text{-}$
pentacarbonyldiiron (Fe—Fe)

17.

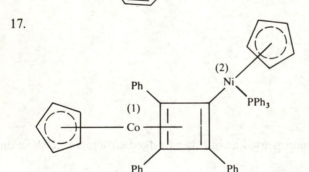

bis$[1,2(\eta^5)\text{-cyclopentadienyl}]\text{-}$
$[\mu\text{-}2,3,4\text{-triphenyl-}2\kappa C^1\text{:}1(\eta^4)\text{-}$
cyclobutadienyl](triphenyl-
phosphine-$2\kappa P$)cobaltnickel

I-10.9.3   Metallocenes [bis(η^5-cyclopentadienyl) entities]

The traditional names ferrocene, manganocene, ruthenocene, nickelocene, etc., are given to the respective bis(η^5-cyclopentadienyl)metal complexes. Such names should be restricted to bis(η^5-cyclopentadienyl)compounds, and not extended to η^6-benzene or η^8-cyclooctatetraene derivatives or other analogues.

Metallocene derivatives may be named by either standard organic suffix (functional) nomenclature or by prefix nomenclature. For a discussion of organic functional suffixes see the *Nomenclature of Organic Chemistry*, 1979 edition, Rule C-22, p. 112.

Note 10q. This unusual placement of the symbol κ preserves the simplified bridging notation.

Substituents are given lowest numerical locants in the usual manner on the equivalent cyclopentadienyl rings of the metallocene entity. The first ring is numbered 1 to 5 and the second ring is numbered 1' to 5'. In entities containing a larger number of cyclopentadienyl groups, the rings are further numbered 1" to 5", 1''' to 5''', etc. The substituent group names -ocenyl, -ocenediyl, -ocenetriyl, etc., are used.

Examples:

1. $[Fe(\eta^5\text{-}C_5H_5)_2)$ ferrocene

2. $[Os(\eta^5\text{-}C_5H_5)_2]$ osmocene

3. $[Ni(CH_3\text{-}\eta^5\text{-}C_5H_4)_2]$ 1,1'-dimethylnickelocene

4. $[Os(\eta^5\text{-}C_5H_5)(HOCH_2CH_2\text{-}\eta^5\text{-}C_5H_4)]$ 2-osmocenylethanol, or (2-hydroxyethyl)osmocene

5. $[Os(\eta^5\text{-}C_5H_5)(CH_3CO\text{-}\eta^5\text{-}C_5H_4)]$ methyl osmocenyl ketone, or acetylosmocene

6. $[Fe(\eta^5\text{-}C_5H_4CH_2CH_2CH_2\text{-}\eta^5\text{-}C_5H_4)]$ 1,3-(1,1'-ferrocenediyl)propane, or 1,1'-trimethyleneferrocene

7. $[Fe_2(\mu\text{-}\eta^5\text{-}C_5H_4CH_2CH_2\text{-}\eta^5\text{-}C_5H_4)(\eta^5\text{-}C_5H_5)_2]$ 1,1"-(1,2-ethanediyl)diferrocene

8. $[Fe(\eta^5\text{-}C_5H_5)\{\eta^5\text{-}C_5H_4As(C_6H_5)_2\}]$ ferrocenyldiphenylarsine, or (diphenylarsino)ferrocene

9. $[Fe(\eta^5\text{-}C_5H_5)_2][BF_4]$ ferrocenium tetrafluoroborate$(1-)$ (Note 10r)

10. $[Fe\{(CH_3)_5\text{-}\eta^5\text{-}C_5\}_2]Cl_2$ decamethylferrocenium$(2+)$ chloride (Note 10r)

11.   [1.1]ferrocenophane (Note 10s)

Note 10r. Ferrocenium is the traditional name for cations derived from ferrocene by the loss of one or two electrons. As such, the -ium ending does not carry the usual meaning that it has in substitutive nomenclature, i.e., the addition of a hydron to a neutral parent compound. The names bis(η^5-cyclopentadienyl)iron$(1+)$ and bis(η^5-cyclopentadienyl)iron$(2+)$ avoid this nomenclature anomaly.

Note 10s. The 'phane' names are based on a proposal by F. Vögtle and P. Neumann, *Tetrahedron*, **26**, 5847 (1970).

12.   [2.2.2]ferrocenophane (Note 10s)

I-10.10   FINAL REMARKS

Although detailed treatments are presented here for much of the nomenclature of coordination chemistry and fundamental matters are given adequate consideration, a variety of pertinent subjects remains open to further development. Organometallic nomenclature in general is the subject of ongoing efforts. The broad subject of stereochemistry includes active areas such as assessment of numbers of diastereoisomers, mathematical models for diastereoisomerism in all polyhedra structures, both mononuclear and polynuclear, and generality of systems for designating chirality. The problem of adapting ligand nomenclature to coordination names becomes more difficult as the complexity of the ligand increases, and developments will continue in this area.

I-11   Boron Hydrides and Related Compounds

CONTENTS

I-11.1   INTRODUCTION

The boron compounds sometimes referred to as 'electron-deficient' include structures which cannot be readily dealt with by any of the classical concepts and procedures of organic or inorganic nomenclature founded on assumptions concerning the localization of bonding electrons. The term 'electron-deficient' implies that there are other 'electron-precise' boron compounds which are in some degree more normal. However, since both classes of compound have bonding molecular orbitals which are precisely filled in all cases, it is recommended that this usage be replaced by 'cluster compounds of boron' or 'polyboron hydrides', as appropriate.

Nomenclature for polyhedral boron hydrides (boranes) and related compounds has presented many problems due to the range of bonding, substitution, and connectivity patterns observed. The understanding of these clusters has done much to rationalize the chemistry of many recently reported inorganic, organometallic, and transition-metal cluster compounds, although there is, as yet, no accepted common nomenclature covering these areas of chemistry. That presented here will be limited to simpler systems of relatively high symmetry; discussion of more complex systems will be reserved for a later publication.

For oxoacids of boron, metal borides, and coordination compounds of boron, see Chapters I-9, I-5, and I-10, respectively.

I-11.2   BORON HYDRIDE NOMENCLATURE

I-11.2.1   Aspects of structures of boron hydrides requiring special consideration

The features of polyboron hydride structures listed below contribute to the complexity of the nomenclature problem.

(i) Connectivity: for many non-metallic elements, the bonds to nearest neighbour-atoms can be rationalized as electron-pair bonds, but for such polyboron-hydride cluster compounds each boron atom can contribute a maximum of 3 electrons to its nearest neighbours, which may number up to 5, 6 or occasionally 7. The connections cannot therefore be regarded as simple electron-pair bonds, and structures of many boron compounds cannot be depicted in terms of straight lines between atoms representing electron-pairs, even though the bonding involved may be strongly directional.

(ii) Triangular association of boron atoms: in many borane structures, clustering in triangular formations occurs, resulting in polyhedra featuring inter-connecting triangular faces. A regular icosahedron, showing in practice the highest symmetry observed, has 20 equilateral triangular faces, 12 vertices, and 30 edges connecting the atoms at the apices. Almost all structures have boron skeletons which may be considered as fragments of an icosahedron or other fully triangulated (closed) polyhedron.

(iii) Hydrogen bridges: polyhedral structures which are fragments of larger polyhedra often have pairs of boron atoms bridged by single hydrogen atoms via three-centre, two-electron bonds.

(iv) Three-centre bonds involving boron atoms: bonding in polyhedral structures with equilateral triangular faces can often be rationalized by depicting three-centre, two-electron bonds between some of the boron atoms.

Representations commonly used for the assemblies discussed in (iii) and (iv) are shown below.

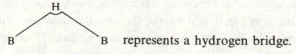

B   B   represents a hydrogen bridge.

represents a three-centre
bond between boron atoms.

For all but the very simplest of polyboron hydrides, several canonical forms can be drawn, which implies bond delocalization (cf., benzene). For example, bonding for the geometrical structure of B_5H_9 (see Section I-11.3.2.2, Example 9) can be represented in plane projection by the four canonical forms represented below.

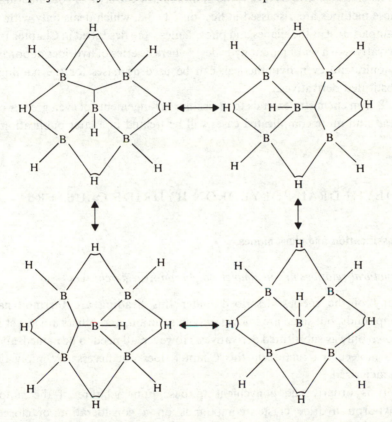

(v) Linkage of polyhedral moieties: this can involve linkage by a bond between boron atoms or the sharing of a common polyhedral vertex, edge, face, or system of faces. Polyboron hydride clusters containing such features have been referred to as '*conjuncto-boranes*' [S. K. Boocock, N. N. Greenwood, J. D. Kennedy, W. S. McDonald, and J. Staves, *J. Chem. Soc., Dalton Trans.*, 790 (1980)].

(vi) Skeletal replacement (subrogation): boron atoms (together with any attached hydrogen atoms) of the polyhedral frameworks can be replaced by many other elements, including metal atoms which may bear ligands.

I-11.2.2 Bases of boron hydride nomenclature

The hydrides of boron are more numerous than those of any other element except carbon, and their nomenclature may be approached from various standpoints, some of which are presented below.

(i) A stoichiometric nomenclature, which is based on that used for hydrocarbons and that used for coordination compounds. The principal difference between this system and hydrocarbon nomenclature is that the number of hydrogen atoms must be defined; it cannot be inferred from simple bonding considerations.

(ii) A structural-descriptor based nomenclature, which may be sub-divided into (a) a semi-systematic style, which seeks to achieve distinctiveness by means of characteristic structural prefixes (closo-, arachno-, etc.), and (b) a subtractive style based on the formal removal of skeletal atoms from fully triangulated polyhedral clusters of boron atoms, the names of which, when neutral, are used as parents in the manner familiar in organic nomenclature; for ionic forms, the names follow the style of coordination nomenclature. These methods are discussed in Section I-11.3.2, which deals only with straightforward examples. Just as the silanes and phosphanes were described in Chapter I-7 as substituted derivatives of a set of parent hydrides, so here a series of hydrides of boron, either neutral or ionic, and even hypothetical, can be used as bases for the naming of a range of substituted derivatives.

Boron chemistry extends to molecular arrangements of even greater complexity than these, and more complicated cases will be treated in a later publication (Note 11a).

I-11.3   POLYHEDRAL POLYBORON HYDRIDE CLUSTERS

I-11.3.1   Classification and class names

I-11.3.1.1   *Structural relationships in simpler polyboron hydride clusters*

Many of the species described under this heading are thermodynamically stable compounds, but some are known only in solution, and others are as yet hypothetical or known only as substituted derivatives. However, all need to be named, and consideration of a molecular structure in this Chapter does not necessarily imply that it has been characterized.

It is orderly and convenient to base nomenclature of the simpler polyhedral polyboron hydride cluster compounds on a consideration of closed, triangulated polyhedra (deltahedra) in which boron atoms occupy all the vertices. Each boron atom has an attached hydrogen atom, and the structures for a series $(B_n H_n)^{2-}$ of closed polyhedra are shown in the first column of Table I-11.1. These are known as *closo*-structures, a corruption of 'clovo', derived from 'clovis' (Latin, *clovis*, a cage; Greek, $\kappa\lambda\omega\beta o\varsigma$).

Note 11a. A proposal (R. M. Adams, Paper 29 delivered at the IMEBORON IV Meeting, Salt Lake City, Utah, USA, July, 1979) to base the nomenclature of inorganic boron cluster compounds on a deltahedral description of a so-called *closo*-polyhedral skeleton has not met with popular acceptance and further proposals for structural approaches to this area of nomenclature have recently been made [J. B. Casey, W. J. Evans, and W. H. Powell, *Inorg. Chem.*, **20**, 1333 (1981); **20**, 3556 (1981); **22**, 2228 (1983); **22**, 2236 (1983); **23**, 4132 (1984); R. W. Rudolph, *Acc. Chem. Res.*, **9**, 446 (1974)].

A series of neutral boron hydrides may be regarded as structurally related to these *closo*-dianions $(B_nH_n)^{2-}$ by formal removal of one BH vertex of highest connectivity (see Section I-11.2.1) and the addition of two hydrons (see Section I-3.5.2) and two new hydrogen atoms, so generating a series of neutral polyboron hydrides of general formula B_nH_{n+4}, in which some hydrogen atoms are linked to a single boron atom and other hydrogen atoms occupy bridging positions between two adjacent boron atoms. The members of this series are known as the *nido*-boranes (Latin, nidus, a nest) (see Section I-11.3.2.2, Example 9) (Note 11b). The neutral *nido*-boranes can lose a hydron producing the monoanions also listed in Table I-11.1. Replacement of skeletal boron atoms by carbon atoms with concomitant adjustment of the number of hydrogen atoms gives a series of carbaboranes with the general formulae shown (see Section I-11.4.3.2, Example 3).

A further series of neutral boron hydrides is likewise structurally related to the *closo*-anions, by formal removal of the BH vertex with the highest connectivity (which would give rise to the *nido*-structure) and one of the adjacent BH vertices. Addition of two hydrons and four hydrogen atoms gives the series of neutral polyboron hydrides of general formula B_nH_{n+6}. The members of this series are known as *arachno*-boranes (Greek, $\alpha\rho\alpha\chi\nu\eta$, a spider's web) (Table I-11.1) (see Section I-11.3.2.2, Example 10).

Extension of this kind of subtractive process generates the *hypho*-boranes (Greek, $\upsilon\phi\epsilon\iota\nu$, to weave), of general formula B_nH_{n+8}, in which the borons occupy the n-vertices of an $n+3$-vertex *closo*-polyhedron, and the *klado*-boranes (Greek, $\kappa\lambda\alpha\delta o\varsigma$, a branch), of general formula B_nH_{n+10}, in which n vertices of an $n+4$-vertex *closo*-polyhedron are occupied. Members of the *hypho*- and *klado*-series have so far been identified only as derivatives.

The structural relationships of all the classes mentioned above are summarized in Table I-11.2. For all these classes, the general formulae are subject to the following limiting provision:

$$\text{number of hydrogen atoms/number of boron atoms} \leq 3$$

I-11.3.1.2   *More complex structural classes*

In addition to those polyboron hydride structures described above, others can occur. For example, any triangular face of a polyhedral polyboron hydride cluster may be capped, which means that an additional skeletal atom is positioned so as to associate directly with all three atoms of an external face to give an extra vertex. This occurs frequently with metallaboron hydride clusters. Examples 1 and 2 on p. 214 illustrate the topography of this class of compound, but their nomenclature is not considered in this Chapter.

Note 11b. *nido*- has been used to refer to other open-structure polyboron hydrides (those with non-triangular faces), but this is not recommended.

Table I-11.1 # Summary of polyhedral structural types in relation to stoichiometry and electron-counting. Arrows indicate principal axes of rotation.

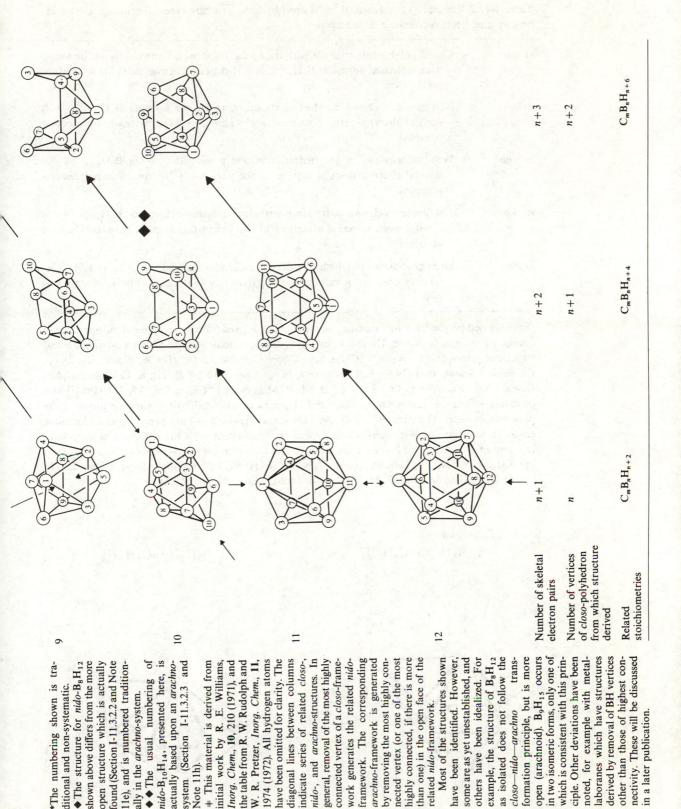

			$n+3$	
				$n+2$
	$n+2$			
	$n+1$			
	$n+1$			
	n			

Number of skeletal electron pairs		
Number of vertices of *closo*-polyhedron from which structure derived		
Related stoichiometries	$C_mB_nH_{n+2}$ $C_mB_nH_{n+4}$ $C_mB_nH_{n+6}$	

9

*The numbering shown is traditional and non-systematic.
◆ The structure for *nido*-B_8H_{12} shown above differs from the more open structure which is actually found (Section I-11.3.2.2 and Note 11e), and is numbered traditionally in the *arachno*-system.
◆ ◆ The usual numbering of *nido*-$B_{10}H_{14}$, presented here, is actually based upon an *arachno*-system (Section I-11.3.2.3 and Note 11h).

10

‡ This material is derived from initial work by R. E. Williams, *Inorg. Chem.*, **10**, 210 (1971), and the table from R. W. Rudolph and W. R. Pretzer, *Inorg. Chem.*, **11**, 1974 (1972). All hydrogen atoms have been omitted for clarity. The diagonal lines between columns indicate series of related *closo*-, *nido*-, and *arachno*-structures. In general, removal of the most highly connected vertex of a *closo*-framework generates the related *nido*-framework. The corresponding *arachno*-framework is generated by removing the most highly connected vertex (or one of the most highly connected, if there is more than one) in the open face of the related *nido*-framework.

11

Most of the structures shown have been identified. However, some are as yet unestablished, and others have been idealized. For example, the structure of B_8H_{12} as isolated does not follow the *closo*—*nido*—*arachno* transformation principle, but is more open (arachnoid). B_9H_{15} occurs in two isomeric forms, only one of which is consistent with this principle. Other deviations have been noted, for example with metallaboranes which have structures derived by removal of BH vertices other than those of highest connectivity. These will be discussed in a later publication.

12

213

Table I-11.2 Summary of polyhedral polyboron-hydride structure-types, according to stoichiometry and electron-counting relationships[a]

closo-	Closed polyhedral structure with triangular faces only; known only as the anion with molecular formula $(B_nH_n)^{2-}$; $(n+1)$ skeletal electron-pairs for an *n*-atom polyhedron.
nido-	Nest-like, non-closed polyhedral structure; molecular formula B_nH_{n+4}; $(n+2)$ skeletal electron-pairs; *n* vertices of the parent $(n+1)$-atom *closo*-polyhedron occupied.
arachno-	Web-like, non-closed polyhedral structure; molecular formula B_nH_{n+6}; $(n+3)$ skeletal electron-pairs; *n* vertices of the parent $(n+2)$-atom *closo*-polyhedron occupied.
hypho-	Net-like, non-closed polyhedral structure; molecular formula B_nH_{n+8}; $(n+4)$ skeletal electron-pairs; *n* vertices of the parent $(n+3)$-atom *closo*-polyhedron occupied.
klado-	Open branch-like polyhedral structure; molecular formula B_nH_{n+10}; $(n+5)$ skeletal electron-pairs; *n* vertices of the parent $(n+4)$-atom *closo*-polyhedron occupied.

[a] Some closed polyhedral structures may be regarded as derived from the capping of the open face of a *nido-* or *arachno*-structure. These do not satisfy the commonly accepted electron-counting and structural principles [K. Wade, *J. Chem. Soc., Chem. Commun.*, 791 (1972); K. Wade, *Adv. Inorg. Chem. Radiochem.*, **18**, 1 (1976); R. E. Williams, *Inorg. Chem.*, **10**, 210 (1971); R. E. Williams, *Adv. Inorg. Chem. Radiochem.*, **18**, 67 (1976); D. M. P. Mingos, *Acc. Chem. Res.*, **17**, 311 (1984)] and probably reflect the number of orbitals and electrons contributed by the capping atom to the polyhedral cluster. The terms *iso-closo-*, *pre-closo-*, and *hyper-closo-* have been suggested for these cases. In addition, some clusters show structures inconsistent with the principle of successive removal of BH vertices of highest connectivity. The terms *iso-nido-*, *iso-arachno-*, etc., have been applied in such cases [R. T. Baker, *Inorg. Chem.*, **25**, 109 (1986); J. D. Kennedy, *Inorg. Chem.*, **25**, 111 (1986); R. L. Johnston and D. M. P. Mingos, *Inorg. Chem.*, **25**, 3321 (1986)].

Examples:

1. $B_4H_4[Co(C_5H_5)]_3$ 2. $B_5H_5[Co(C_5H_5)]_3$

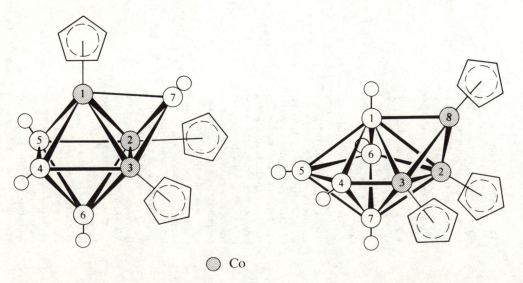

⬡ Co

Linkage between polyhedral boron hydride clusters is covered by the general term 'conjuncto-boranes'. The types of linkage considered here fall into several general classes. (a) Linkage by a direct two-centre boron–boron bond between different clusters (whether closo- or not), with corresponding elimination of a hydrogen atom from each, for example, this is shown in Examples 3, 4, and 5 which display three different possible isomers for the association of two ions $(B_{10}H_{10})^{2-}$ to give $(B_{20}H_{18})^{4-}$.

Examples:

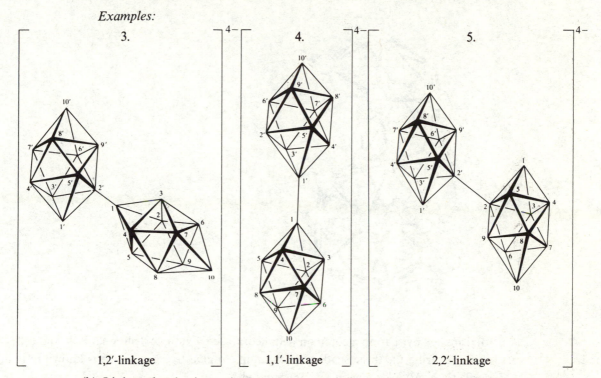

3.   4.   5.

1,2'-linkage   1,1'-linkage   2,2'-linkage

(b) Linkage by sharing a single vertex, termed the *commo*-vertex, as in Example 6, which is a structure involving the sharing of a boron atom (shaded) between a B_7- and B_9-cluster.

Example:

6. $B_{15}H_{23}$

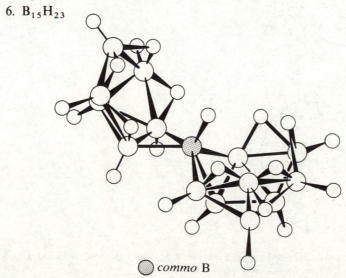

⬤ *commo* B

(c) Linkage by sharing a polyhedron-edge made up of two linked vertices. In Example 7 two B_{10}-clusters have been linked by sharing the edge indicated by shaded atoms to give the centro-symmetric neutral hydride *anti*-$B_{18}H_{22}$ (Note 11c).

Example:
7. $B_{18}H_{22}$

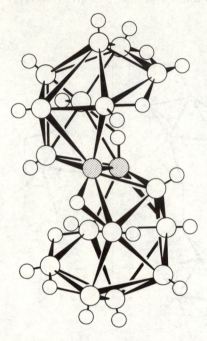

(d) Linkage by sharing a common deltahedral face. Example 8 shows a B_{11}- and a B_{12}-cluster sharing the three shaded boron atoms to give $B_{20}H_{18}$, which is known only in derivatives. All hydrogen atoms are omitted.

Example:
8. $B_{20}H_{18}$

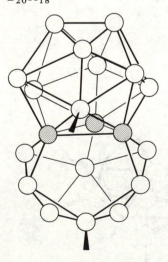

Note 11c. For the definition of *anti*- and *syn*-, see Y. M. Cheek, N. N. Greenwood, J. D. Kennedy, and W. S. McDonald, *J. Chem. Soc., Chem. Commun.*, 80 (1982).

Each free valence position shown is a site of attachment of a neutral organic ligand in one of the known examples.

(e) More extensive sharing between clusters. Although the *closo*-borane in Example 9 may be named as a fully triangulated polyhedral structure, it may also be regarded as sharing the four shaded boron atoms between two B_{12}-clusters. The hydrogen atom on each of the non-shaded boron atoms has been omitted for clarity.

Example:

9. *closo*-$B_{20}H_{16}$

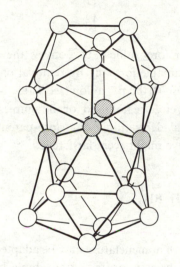

I-11.3.2   Naming of individual hydrides of boron

I-11.3.2.1   *Stoichiometric names*

Neutral polyboron hydrides are called boranes and the simplest possible parent structure, BH_3, is given the name 'borane' (Note 11d).

For higher boranes, the number of boron atoms in the molecule is indicated by an appropriate numerical prefix di-, tri-, tetra-, penta-, etc., added to the stem 'borane'. The latin nona- and undeca- are used in preference to the Greek ennea- and hendeca- for conformity with hydrocarbon nomenclature (see the *Nomenclature of Organic Chemistry*, 1979 edition, p. 5).

The number of hydrogen atoms in the molecule is indicated by enclosing the appropriate arabic numeral in parentheses directly following the name. Thus B_2H_6 is named diborane(6), B_6H_{10} is named hexaborane(10), and $B_{10}H_{14}$ may be named decaborane(14).

Such names convey that a cluster of x boron atoms is associated with y hydrogen atoms, but give no direct information as regards structure.

Note 11d. The older nomenclature using 'borine' has been abandoned.

I-11.3.2.2   *Structural descriptor names* (*general*)

The usual names for the simplest neutral boron hydrides are given below, with reference to representations of their structural formulae.

			Example number in *Section I-11.3.2.2*
(a)	B_4H_{10}	*arachno*-tetraborane(10)	8
(b)	B_5H_9	*nido*-pentaborane(9)	9
(c)	B_5H_{11}	*arachno*-pentaborane(11)	10
(d)	B_6H_{10}	*nido*-hexaborane(10)	11
(e)	$B_{10}H_{14}$	*nido*-decaborane(14)	12

The name *nido*-hexaborane(10) therefore suffices to convey the skeletal boron arrangement shown in Example 11, obtainable formally by removal of one of the two highest-connectivity vertices from the *closo*-$(B_7H_7)^{2-}$ anion (see Table I-11.1). It should be noted that the prefixes *nido*-, *arachno*-, etc., are not used for the simplest boranes (such as Examples 1 and 7) for which formal derivation from *closo*-parent structures by successive subtractions as described above might seem far-fetched.

Example:

1. tetraborane(6), $H_2B-BH-BH-BH_2$
   or *catena*-tetraborane(6)

The procedures of organic hydrocarbon nomenclature may be adapted for the names of simple boron parent-hydrides, whether in chains or rings, for example, by basing names for branched chains on the senior unbranched chain of boron atoms together with the arabic numeral in parentheses to indicate the number of hydrogen atoms, using the suffix -ene to denote a double bond. In the case of ring-compounds, the prefix *cyclo*- or the extended Hantzsch-Widman nomenclature system may be used.

Examples:

2. $H_2B-BH-BH_2$ triborane(5), or
                        catena-triborane(5)

3. $HB=B-BH_2$ triborene(3), or
                        catena-triborene(3)

4.   *cyclo*-tetraborane(4),
                        or tetraboretane (the
                        Hantzsch-Widman name)

5.   2-boryltriborane(5)

In Example 5 the alkane-nomenclature principle of basing the name of a branched acyclic compound on that of the longest unbranched chain present has been followed, and the numeral in parentheses refers to the hydrogen count of the unsubstituted parent chain. The known compound shown in Example 6 is thus named as a derivative of triborane(5).

Examples:

6. 2-(difluoroboryl)-1,1,3,3-tetrafluorotriborane(5)

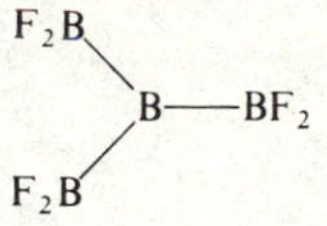

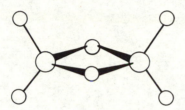

7. diborane(6), B_2H_6

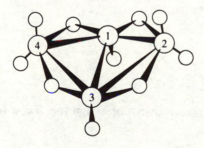

8. *arachno*-tetraborane(10), B_4H_{10}

9. *nido*-pentaborane(9), B_5H_9

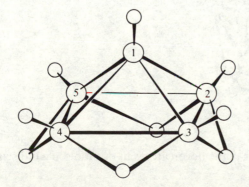

10. *arachno*-pentaborane(11), B_5H_{11}

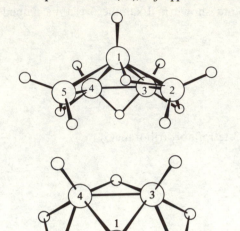

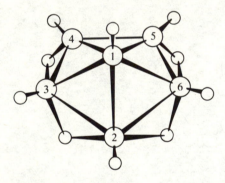

11. *nido*-hexaborane(10), B_6H_{10}

The traditional numbering scheme shown does not assign the lowest locant set to the bridging hydrogen atoms.

12. *nido*-decaborane(14), $B_{10}H_{14}$

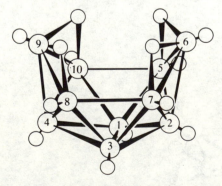

The non-standard numbering of the boron atoms in this molecule is discussed in Section I-11.3.2.3, Note 11h.

13. *nido*-octaborane(12), B_8H_{12} (Note 11e)

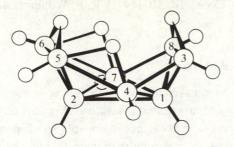

14. *arachno*-nonaborane(15), B_9H_{15}

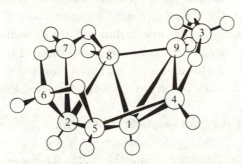

The stoichiometric names of polyhedral boranes and their derivatives are often modified by the descriptive prefixes *closo*-, *nido*-, etc., discussed in Section I-11.3.1.1, according to whether the polyhedral polyboron hydride cluster is a fully triangulated (closed) polyhedron or has one or more vertices unoccupied. Systematic relationships among (a) the open structural types represented by the prefixes *nido*-, *arachno*-, etc., (b) the series of *closo*-polyhedra, and (c) electron-counting rules are well recognized (Note 11f). However, exceptions to these rules do occur even in simple polyboranes (e.g., B_8H_{12} in Section I-11.3.2.2, Example 13) and certainly with more complex structures incorporating capping of polyhedral faces. For this reason these prefixes must be considered optional and should be used only to qualify (i) structures based on polyhedra which are readily rationalized by their relationship to the series of closed polyhedra, and (ii) molecular formulae as summarized in Table I-11.2, and (iii) particular electron counts arrived at by application of electron-counting rules.

The 6- and 12-vertex *closo*-structures of Table I-11.1, have such symmetry that their vertices are all identical and the same open structure, whether *nido*- or *arachno*-, is obtained, regardless of whichever vertex or vertex-pair, respectively, is removed. However, in the case of the other structures shown, the prefixes *nido*- and *arachno*- are not sufficiently precise to convey a unique structure unless the *nido*-prefix is taken to convey that a vertex of maximum skeletal connectivity has been removed, and *arachno*- that another such vertex has been removed from the open face of a *nido*-structure as already so defined. However, to generate the B_7- and B_8-*arachno*-structures shown, a further choice has to be made from more than one vertex of equally high connectivity in the related *nido*-structure. The selection of the particular vertex removed has been rationalized in terms of the greater stability of the resulting structures which contain the lesser number of

Note 11e. This cluster, which possesses a *nido*-stoichiometry, adopts an alternative open *nido*-structure which may also be considered to be consistent with classification as an *arachno*-structure (cf., Table I-11.1).

Note 11f. See references in Note 11a and Table I-11.2.

skeletally two-coordinate vertices, referred to in the original publications as 'leaving BH_3 groups' [R. E. Williams, *Inorg. Chem.*, **10**, 210 (1971); R. E. Williams, *Adv. Inorg. Chem. Radiochem.*, **18**, 67 (1976)].

I-11.3.2.3   *Systematic numbering of polyhedral clusters*

It is necessary to number the boron skeleton for each cluster systematically, so as to permit the unambiguous naming of substituted derivatives. For this purpose, the boron atoms of *closo*-structures are considered as occupying planes disposed sequentially, perpendicular to the axis of highest-order symmetry (if there are two, the 'longer', in terms of the greater number of perpendicular planes crossed, is chosen). Numbering begins at the nearest boron atom when the cluster is viewed along this axis. It proceeds either clockwise or counter-clockwise (Note 11g), first dealing with the skeletal atoms of the next plane encountered, then continuing in the same sense with those of the next beginning with the boron atom nearest to the lowest-numbered boron atom in the preceding plane, and so on as necessary, until the distal boron atom or edge is reached. For an example, see the numbered structure $(B_{10}H_{10})^{2-}$ in Table I-11.1.

For symmetry reasons, it may not be possible to assign a unique vertex to start numbering. Thus, in the B_6- and B_{12}-*closo*-clusters of Table I-11.1, column 1, any position may equally be taken as number 1. Similarly, in the B_7-*closo*-polyhedron either the uppermost or lowest positions, as drawn, could be taken as 1. Thereafter, however, numbering proceeds as described.

The numbering of a *nido*-cluster is derived from that of the related *closo*-structure. Note that the boron atom formally removed from the *closo*-cluster to generate the *nido*-structure has the highest possible connectivity and number. That number may not be the highest locant generated in the closo-structure (see, for example, B_7H_{11} and B_9H_{13} in Table 1-11.1). (Note 11h).

In the case of *arachno*- and of more open clusters, the opened side is presented towards the observer and the boron atoms considered as projected onto a plane at the rear. They are then numbered sequentially in zones, commencing at the central boron atom of highest connectivity and proceeding clockwise or anti-clockwise until the innermost zone is completed in the same sense from the 12 o'clock position around the next zone, and so on until the outermost zone is completed (Note 11i). When there is a choice, the molecule is so oriented that the 12 o'clock position lies in a position decided by sequential application of the following criteria.

(a) The 12 o'clock position lies in a symmetry-plane which contains as few boron atoms as possible.

(b) The 12 o'clock position lies in that portion of the symmetry-plane which contains the greatest number of skeletal atoms (B_5H_{11}, Section I-11.3.2.2, Example 10).

(c) The 12 o'clock position lies opposite the greater number of bridging atoms (B_9H_{15}, Section I-11.3.2.2, Example 14). Criteria (a)–(c) may fail to effect a decision, and where a

Note 11g. If these two directions give a different locant-set for replacement atoms and substituents, the lower locant-set at the first point of difference is selected.

Note 11h. One exception is *nido*-$B_{10}H_{14}$ which, by established usage, has a numbering based on that for *arachno*-systems.

Note 11i. This treatment means that the numbering of a *closo*-parent is unlikely to carry over into the corresponding *arachno*-parent.

symmetry plane is lacking, they are inapplicable. In such cases traditional numbering is used, Example 1.

Example:

1. *arachno*-hexaborane(12), B_6H_{12}

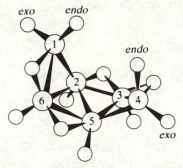

Where two clusters are connected, their numbering follows these same principles for the individual components except that the numerals for the minor cluster are primed, points of attachment being given the lowest numeral locants compatible with the fixed numbering of the polyhedra.

In the case of *commo*-linkage, each cluster is numbered separately. The minor cluster is again distinguished in numbering by primes. Consequently any *commo*-atoms receive two numbers, one each from the major and minor cluster.

If the two connected clusters are identical, assignment of major/minor status will depend on the pattern of replacement (subrogation) and then, if still unclear, on that of the substitution. Thus, in Section I-11.3.1.2 Examples 3, 4, and 5, if the uppermost vertex of the higher cluster of the pairs were occupied by a carbon atom, this cluster would be accorded major status. A substituted moiety, all other things being equal, takes precedence over an unsubstituted moiety. This kind of assignment will be discussed in detail in a later publication.

I-11.3.2.4   *Systematic naming giving hydrogen-atom distribution*

Once skeletal numbering has been assigned, it is possible to devise names giving the precise hydrogen-atom distribution. In the open boranes each boron atom can be assumed to carry at least a single hydrogen atom as in the *closo*-parent, but it is also necessary to assign the positions of the bridging hydrogen atoms. This can be achieved by adapting the principle of indicated hydrogen (Note 11j) and using the symbol μ (see Chapter I-10), preceded by the locants for the skeletal positions so bridged in ascending numerical order. Note that the designator *H* is used for a hydrogen atom, not the name 'hydro' which could be taken to imply the presence of hydrogen atoms in addition to any shown by the normal methods. The punctuation should be compared with practice cited in Chapter I-10.

Example:

2. B_9H_{15}
   3,4:3,9:5,6:6,7:7,8-penta-*μH*-*arachno*-nonaborane(15)

Note 11j. See the *Nomenclature of Organic Chemistry*, 1979 edition, Rule A-21.6, p. 215.

This name does not convey the position of the additional hydrogen atom on boron atom (3) of B_9H_{15}, the structure of which is shown in Example 14, Section I-11.3.2.2. Where the distribution of bridging hydrogen atoms is unambiguous (for certain *nido-* and *arachno-* boranes theory predicts only one structure), then the bridging descriptors and locants may be omitted. An alternative coordination name which gives the complete hydrogen-atom distribution is shown below, Example 3 (Note 11k).

Example:

3. B_9H_{15}

   3,4:3,9:5,6:6,7:7,8-penta-μ-hydro-1,2,3,3,4,5,6,7,8,9-decahydro-*arachno*-nonaboron

I-11.3.2.5   *Steric descriptors* endo- *and* exo-

Boron atoms in open clusters (*arachno-*, etc.) may carry two terminal hydrogen atoms, one corresponding to the externally directed B–H bonds of the *closo*-polyhedron and the other directed tangentially to the open concavity of the cluster. Where necessary, the former may be distinguished by the designator *exo-*, and the latter by the designator *endo-*.

The *endo*-hydrogen atoms are designated in Section I-11.4.2, Example 1, in which the further hydrogen atoms on B_3 and B_4 are obviously *exo*-hydrogen atoms. See also Sections I-11.3.2.2, Example 10, and I-11.3.2.3, Example 1. Compounds carrying groups other than hydrogen atoms in *endo*-positions have been characterized.

I-11.3.2.6   *Structural isomers*

There is no simple general method for distinguishing structural isomers among the boron hydrides, but the principles expounded above should allow the assignment of a specific name to each isomer. The prefixes *iso-* and *neo-* have each been used to distinguish structural isomers, but their use should be reserved for isomers whose structures have not been determined. They have no specific structural significance. However, a method which will describe all known isomeric polyhedral systems has been proposed (see references in Note 11a).

I-11.3.2.7   *The 'debor' method for naming open polyboron hydride clusters*

Another method of naming such structures is to systematize the use of *nido-* and *arachno-* by designating precisely the removal-sites from *closo*-parents by means of numerical locants. This subtractive approach has been generalized in the so-called 'debor' method. In these names, each of the BH-vertices removed from the corresponding *closo*-parent structure is denoted by means of the prefix debor-, together with the numerical locant (see below) for the removed vertex. The meaning of debor- here is analogous to that of nor- in subtractive organic nomenclature which conveys the removal of a CH_2 group from a named structure (Note 11l). Removal of more than one vertex is conveyed in the name by didebor-, tridebor-, etc. (cf., dinor-, trinor-). The name of the *closo*-parent then follows.

Note 11k. For use with boron compounds, 'hydro' is exceptionally preferred to the more general 'hydrido' for the hydrogen ligand.

Note 11l. See the *Nomenclature of Organic Chemistry*, 1979 edition, Rules C-42 and 43, p. 116, and Section F-4.

The debor locant should be the highest number available consistent with the fixed numbering of the parent closed polyhedron. The number of hydrogen atoms in the open hydride is indicated by adding the appropriate arabic numeral in parentheses at the end of the name.

Example:

1. *nido*-hexaborane(10), B_6H_{10}, can be named 7-debor-*closo*-heptaborane(10).

The distribution of bridging atoms is not automatically conveyed by such names and, in cases where there is doubt, this can be resolved, as in Section I-11.3.2.3, by means of μH with appropriate locants.

Example:

2. *nido*-hexaborane(10) may also be designated 2,3:2,6:3,4:5,6-tetra-μH-7-debor-*closo*-heptaborane (10).

Whichever method is used, the numeral in parentheses at the end of the name represents the number of hydrogen atoms in the named hydride and not in the *closo*-parent structure.

The methods presented above have advantages in different specific circumstances, and the use of a given method should be determined by the requirements of the user.

I-11.4   SUBSTITUTION AND REPLACEMENT IN BORON CLUSTERS

I-11.4.1   Hydrogen substitution

Compounds containing a single boron atom are named as derivatives of BH_3, borane. The procedures of organic substitutive nomenclature (see Section I-7.2.3) are followed to name substituted derivatives of neutral boron hydrides. The number of hydrogen atoms of the parent hydride is conveyed by the appropriate hydride name (ending in -ane), together with appropriate hydrogen designators. Substitution is then conveyed by citing the names of the substituting groups in alphabetical order, qualified where applicable by numerical prefixes, di-, tri-, etc., and each radical being assigned its numerical positional locant. Numerals in parentheses at the end of a name indicate the hydrogen atom population prior to substitution.

Examples:

1. $HBCl_2$ dichloroborane
2. BBr_2F dibromofluoroborane
3. $B(OH)_3$ trihydroxyborane [also named boric acid (see Table I-9.1)]
4. $BCl(OCH_3)_2$ chlorodimethoxyborane
5. $B(OC_2H_5)_3$ triethoxyborane [also named triethyl borate (see Chapter I-9)]
6. $B(CH_3)(SCH_2CH_3)_2$ bis(ethylthio)(methyl)borane(Note 11m)

Note 11m. These compounds have also been named as derivatives of borinic (R_2BOH) and boronic [$RB(OH)_2$] acids, but such names are no longer recommended by IUPAC.

7. $BCl(OCHCl_2)_2$ chlorobis(dichloromethoxy)borane (Note 11m)
8. $B[NHN(CH_3)_2]_3$ tris(2,2-dimethylhydrazino)borane
9. $B(OMe)Me_2$ methoxydimethylborane (Note 11m)
10. $BMe(OH)_2$ dihydroxy(methyl)borane (Note 11m)
11. Br_2BBBr_2 tetrabromodiborane(4)

12.

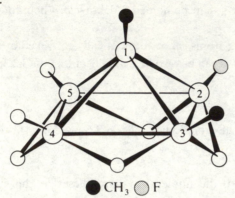

2-fluoro-1,3-dimethylpentaborane(9)

● CH_3 ◍ F

13.

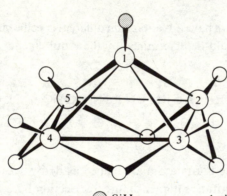

◍ SiH_3 1-silyl-2,3:2,5:3,4:4,5-tetra-μH-pentaborane(9)

Substituents for bridging groups are designated in the name by the prefix μ placed before the name of the substituent group. Again, when there is a choice of locant, the lowest locants are preferred.

Examples:

14.

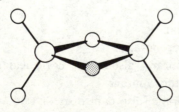

μ-amino-diborane(6)

◍ NH_2

15.

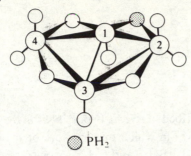

1, 2-μ-phosphino-*arachno*-
tetraborane(10)

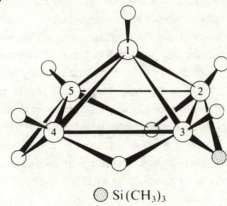

Si(CH$_3$)$_3$

2,3-μ-trimethylsilyl-2,5:3,4:4,5-tri-μH-*nido*-pentaborane(9)

I-11.4.2   Adduct formation

Comparison of any structure in the middle column of Table I-11.1 with that immediately to its right shows a progression to a more open structure, accompanied by the addition of two more hydrogen atoms and, consequently, two further electrons. Adduct formation by a neutral ligand such as triphenylphosphine generally has the analogous effect of feeding additional electrons into the cluster, causing rearrangement to a more open derivative with the same number of hydrogen atoms. Thus the *hypho*-cluster, Example 1, results from adduct-formation with *nido*-pentaborane(9). The compound is currently named as a *quasi*-addition compound.

Example:

1. (trimethylphosphine)—*hypho*-pentaborane (9) (2/1), [B$_5$H$_9$·(PMe$_3$)$_2$]

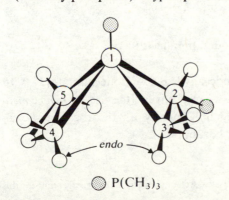

P(CH$_3$)$_3$

In this example, the *endo*-hydrogens are those at the front of the structure pointing downwards and towards the reader.

I-11.4.3   Skeletal replacement (subrogation)

I-11.4.3.1   *General*

In the boron cluster formations described above, it is possible for the essential skeletal structure to be preserved in derivatives in which one or more of the boron atoms are replaced by other atoms (note that BH_2, BH^-, and CH are isoelectronic). The names of such species are formed by an adaptation of organic replacement nomenclature to give carbaboranes, azaboranes, phosphaboranes, thiaboranes, etc.

In the heteroboranes, the number of nearest neighbours to the heteroatom is variable and can be 5, 6, 7, etc. Therefore, in the adaptation of organic replacement nomenclature to polyborane compounds, the replacement of a boron atom by another atom is indicated in the name along with the number of hydrogen atoms in the resulting polyhedral structure. The prefixes *closo-*, *nido-*, *arachno-*, etc., are retained as described above. The positions of the supplanting heteroatoms in the polyhedral network are indicated by locants which are the lowest possible numbers taken as a set consistent with the numbering of the parent polyborane. If a choice remains for locant-assignment within a given set, then priority for low numbering should be assigned to the element encountered first using Table IV in the usual way.

I-11.4.3.2   *Carbaboranes* (Note 11n)

In this important class of compound, the general formula is $[(CH)_a(BH)_m H_b]^c$ where c can be positive, negative, or zero. The CH groups occupy polyhedron vertices and other hydrogen atoms are either bridging (μH) or *endo*-terminal. Carbaborane names are based on the corresponding polyboron skeletal parent hydride, and the prefixes *closo-*, *nido-*, *arachno-*, etc., retain their significance. The locants for the replacing atoms are made as low as is consistent with the fixed polyhedral numbering.

The hydrogen-atom population of the actual compound concerned (and not that of the parent all-boron skeletal compound) is added as an arabic numeral in parentheses at the end of the name. This numeral is retained in names for their derivatives formed by replacement of hydrogen atoms. In all the following diagrams carbon atoms are represented by black circles and boron atoms by white circles.

Example:

1. $C_2 B_{10} H_{12}$, dicarba-*closo*-dodecaborane(12)

This is isoelectronic with dodecahydro-*closo*-dodecaborate(2−). It has three positional isomers: 1,2-, 1,7-, and 1,12-, respectively (the use of *ortho-*, *meta-*, and *para-* for these isomers is not recommended). Similarly, 1,6-dicarba-*closo*-hexaborane(6) is isoelectronic with hexahydro-*closo*-hexaborate(2−).

Note 11n. The contraction 'carboranes' is also well established as the generic name for this class of compound.

Examples:

2. $C_2B_3H_5$, 1,5-dicarba-*closo*-pentaborane(5)

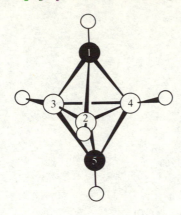

3. $C_2B_4H_8$, 4,5:5,6-di-μH-2,3-dicarba-*nido*-hexaborane(8)

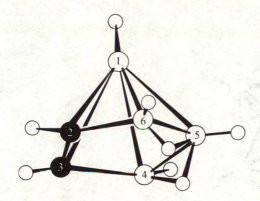

Note that locants for skeletal replacement take precedence over those for hydrogen bridging. The number of bridging hydrogen atoms is usually different for heteroboranes compared with the parent polyboranes, and for numbering purposes only the symmetry of the parent boron skeleton is considered.

Examples:

4. $C_2B_5H_7$, 2,4-dicarba-*closo*-heptaborane(7)

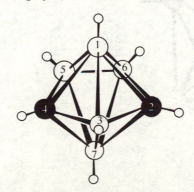

5. $C_2B_6H_8$, 1,7-dicarba-*closo*-octaborane(8)

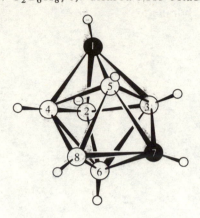

6. $C_2B_8H_{10}$, 1,10-dicarba-*closo*-decaborane(10)

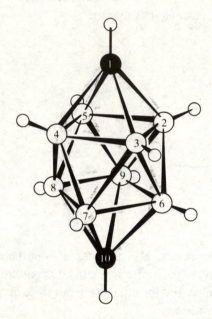

7. $C_2B_8H_{14}$, 6,9-dicarba-*arachno*-decaborane(14)

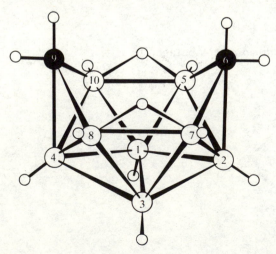

8. $C_6H_{17}B_4Cl_2Si_2$, 2,3-bis(chlorodimethylsilyl)-4,5:5,6-di-μH-2,3-dicarba-*nido*-hexaborane(8)

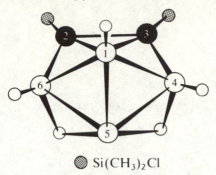

⬡ $Si(CH_3)_2Cl$

I-11.4.3.3   *Metallaboranes and metallacarbaboranes*

These will be treated in more detail in a later publication, but a few examples here will suffice to indicate how the essential polyhedral polyboron hydride cluster morphology can be retained when one or more boron atoms is replaced by a metal. Metallaborane clusters may be neutral or ionic.

Replacement nomenclature is adapted to name such species, citing the appropriate prefixes from Table VI (see also Table I of the *Nomenclature of Organic Chemistry*, 1979 edition, p. 459). Lowest locants *consistent with the fixed polyhedral numbering* are assigned and, where there remains a choice, the preference for lowest locants goes to the elements in the order in which they are first encountered in following the directional arrow of Table IV (see also Section I-11.4.3.1).

Examples:

1. $[B_8H_{12}[Co\{C_5(CH_3)_5\}]_2]$

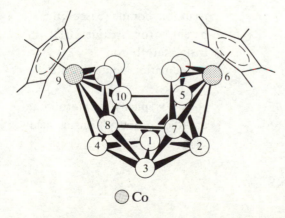

⬡ Co

6,9-bis(η^5-pentamethylcyclopentadienyl)-5,6:6,7:8,9:9,10-tetra-μH-6,9-dicobalta-*nido*-decaborane(12) (Note 11o).

Note 11o. A singly attached hydrogen atom at each boron atom has been omitted for clarity. For a discussion on naming ligands attached to metal ions, see Chapter I-10. In determining the number to be placed in parentheses, it is normally not assumed that the metal atoms in compounds such as these would carry hydrogen atoms in the unsubstituted parent. Hence, any such hydrogen atoms need to be specified.

2. $[(CH_2(BH)_3Fe(CO)_3]$

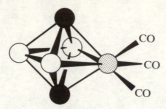

2,2,2-tricarbonyl-1,6-dicarba-2-ferra-*closo*-hexaborane(5)
(not 2,4-dicarba-1-ferra-)

3. $[(CH_2)_2(BH)_3Fe(CO)_3]$

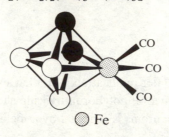

◎ Fe

3,3,3-tricarbonyl-1,2-dicarba-3-ferra-*closo*-hexaborane(5)
(not 1,3-dicarba-2-ferra-)

I-11.4.3.4   *Organoboron compounds*

In general, for compounds with two-centre, covalent bonds which contain non-metals besides boron and hydrogen, names are formed on the basis of the boron moiety (ion, hydride, or cluster name). However, organoboron compounds may also be named in accordance with the rules of organic nomenclature.

Examples:

1. $B(OCOMe)_3$ triacetoxyborane (based on parent boron hydride, BH_3), or tris(acetato)boron (coordination nomenclature)

2. 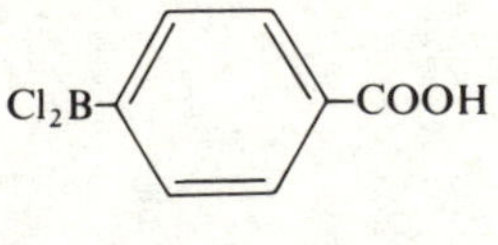 Cl_2B—⟨benzene ring⟩—COOH   (4-carboxyphenyl)dichloroborane, or 4(dichloroboryl)benzoic acid

I-11.5   NAMES FOR IONS

I-11.5.1   Anions

Examples of anions occur in all the structural classes of compound described above. Their names are formed using the same numerical and structural prefixes as those of the neutral hydrides, but the names end in 'borate' instead of 'borane'. The hydrogen-atom population is indicated by an appropriate numerical prefix combined with 'hydro', and the anionic charge is indicated in parentheses at the end. Thus, the structures of Table

I-11.1 with all their hydrogen atoms are named pentahydro-*closo*-pentaborate(2−), hexahydro-*closo*-hexaborate(2−), heptahydro-*closo*-heptaborate(2−), and so on, and Examples 3, 4, and 5, of Section I-11.3.1.2 are octadecahydro-1,1'-(or 1,2'- or 2,2'-) bi-*closo*-decaborate(4−). For open structures this nomenclature may need to be amplified as described in Sections I-11.3.2.3 and I-11.3.2.4 in order to assign locants to the hydrogen atoms.

Examples:

1. $[BH_4]^-$ tetrahydroborate(1−)

2. $[H_3BHBH_3]^-$ heptahydrodiborate(1−), or
   μ-hydro-hexahydrodiborate(1−)

3. $[B_3H_8]^-$ octahydro-*cyclo*-triborate(1−),
   or 1,2:1,3-di-μ-hydro-
   hexahydro-*cyclo*-triborate(1−)

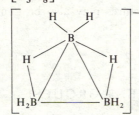

4. $[B_6H_9]^-$ nonahydrohexaborate(1−), or
   2,3:2,6:4,5-tri-μ-hydro-
   hexahydro-*nido*-hexaborate(1−)

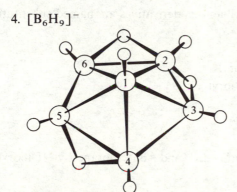

For salts of metals of assumed unequivocal oxidation state the ionic charge may be omitted from the name.

Examples:

5. $Na[BF_4]$ sodium tetrafluoroborate

6. $NH_4[B(C_6H_5)_4]$ ammonium tetraphenylborate

7. $Na_2[H_3BC(O)O]$ sodium carboxylatotrihydroborate(2−),
   or disodium carboxylatotrihydroborate

I-11.5.2   Cations

The names of cations are not given a distinguishing termination; the name of the ion ends in 'boron'. All attached atoms and groups are denoted by ligand names.

Examples:

1. $[BH_2(NH_3)_2]Cl$ diamminedihydroboron(1+) chloride

2. $[BH_2(py)_2]^+$ dihydrobis(pyridine)boron(1+)

3. $[B_{10}H_7(NH_3)_3]^+$ triammineheptahydro-*closo*-decaboron(1+)

4. $[BH_2(NH_3)_2][B_3H_8]$ diamminedihydroboron octahydro-*cyclo*-triborate

I-11.5.3   **Structures containing both cationic and anionic centres (zwitterions)**

These are named as anions, and the prefixes, which are cited in alphabetical order, include an appropriate substitutive cation-prefix.

Examples:

1. $Me_3P^+-CH_2B^-H_3$ trihydro(trimethylphosphoniomethyl)borate

2. 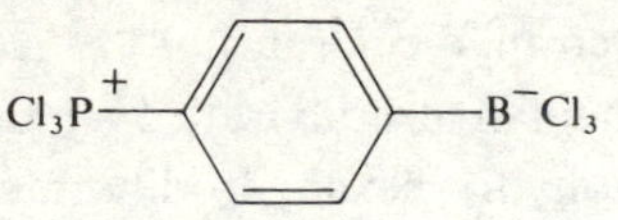   trichloro[4-(trichlorophosphonio)phenyl]borate

Cl_3P^+ —⬡— B^-Cl_3

Where the charge-sites are adjacent, the structures may also be named as addition compounds (see Chapter I-5).

I-11.6   NAMES FOR RADICALS (SUBSTITUENT GROUPS)

The H_2B- group is named 'boryl' and its derivatives are named substitutively.

Examples:

1. Cl_2B- dichloroboryl
2. $(OH)_2B-$ dihydroxyboryl
3. $O=B-$ oxoboryl
4. $S=B-$ thioxoboryl

Polyvalent forms are $HB<$, boranediyl, and $-B<$, boranetriyl. 'Diboryl' means two H_2B- groups.

Example:

5. $BrB<$ bromoboranediyl

If there is no specification of bond number for the radical or group, 'borio' may be used.

Examples:

6. Cl_2B- dichloroborio
7. $HB<$ hydroborio
8. $HOB<$ hydroxoborio
9. $-B<$ borio

For radicals and groups derived from polyboranes, the parent hydride-name, complete with the number of hydrogen atoms in parentheses, is amended as follows below. For univalent groups only, the 'e' of '-borane' is elided. The locants for the free valence or valences, separated where more than one by commas, and flanked by hyphens, precede the suffixes -yl, -diyl, -triyl, etc. When there is a choice in numbering, free-valence positions are given lowest locants consistent with the fixed numbering, but after any substituted (subrogated) positions have been assigned (see Section I-11.4.3).

234

Examples:

10. H_2BH_2BH- diboran(6)-1-yl
11. $-HBH_2BH-$ *cis*-diborane(6)-1,2-diyl
12. $H_2BH_2B<$ diborane(6)-1,1-diyl
13. $-HBBH-$ diborane(4)-1,2-diyl
14. *nido*-pentaboran(9)-1-yl

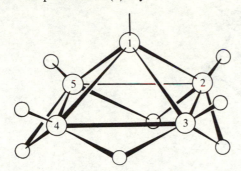

15. *arachno*-tetraborane(10)-2,2-diyl

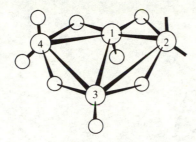

16. 1,1,1,3,3,3-hexacarbonyl-2,5-dicarba-1,3-diferra-*closo*-heptaboran-6-yl

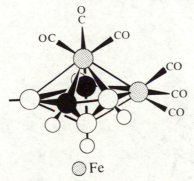

Fe

Radicals or groups formed by the removal of bridging hydrogen atoms are indicated by listing the numerals for the atoms to which the bridge was attached, separated by a comma and enclosed in parentheses.

Examples:

17. 1,2-dicarba-*closo*-dodecaborane(12)-1,2-diyl

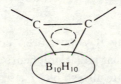

The representational abbreviation of Example 17 for the regular dodecahedral skeleton $B_{10}C_2$ (cf., structure in line 12 of Table I-11.1) is much used.

Examples:

18. 3-chloro-2,3-dicarba-*nido*-hexaborane(8)-2,4-diyl

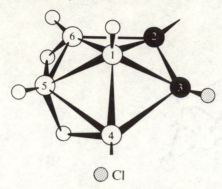

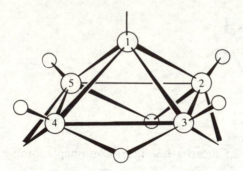

19. *nido*-pentaborane(9)-1,(2,3-μ)(4,5-μ)-triyl

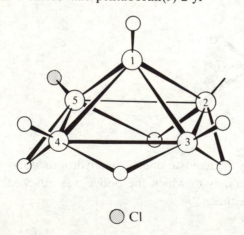

20. 5-chloro-*nido*-pentaboran(9)-2-yl

I-11.7   FINAL REMARKS

The regular *closo*-structures up to and including B_{12} have been considered in this Chapter. However, theory and experiment show that larger polyhedra and other closed polyhedra having faces of more than three edges are also stable. These and other boron structures of greater complexity present particular nomenclature problems and will be considered in a later publication.

The structural variety and complexity touched on in this Chapter are also to be found in metal borides, but they are best treated using stoichiometric names (Chapter I-5).

Interstitial borides and boron addition compounds are treated as in Chapter I-5, some boron ring-compounds are described in Chapter I-7 (Note 11p), and boron oxoanions are discussed in Chapter I-9.

Note 11p. See also the *Nomenclature of Organic Chemistry*, 1979 edition, Rule D-7.5, p. 440.

The Tables

TABLE I

Table I Names, symbols, and atomic numbers of the atoms [elements]

Name	Symbol	Atomic number	Name	Symbol	Atomic number
Actinium	Ac	89	Manganese	Mn	25
Aluminium	Al	13	Mendelevium (Unnilunium)	Md	101
Americium	Am	95	Mercury (Hydrargyrum)	Hg	80
Antimony (Stibium)	Sb	51	Molybdenum	Mo	42
Argon	Ar	18	Neodymium	Nd	60
Arsenic	As	33	Neon	Ne	10
Astatine	At	85	Neptunium	Np	93
Barium	Ba	56	Nickel	Ni	28
Berkelium	Bk	97	Niobium	Nb	41
Beryllium	Be	4	Nitrogen (Azote)	N	7
Bismuth	Bi	83	Nobelium (Unnilbium)	No	102
Boron	B	5	Osmium	Os	76
Bromine	Br	35	Oxygen	O	8
Cadmium	Cd	48	Palladium	Pd	46
Caesium	Cs	55	Phosphorus	P	15
Calcium	Ca	20	Platinum	Pt	78
Californium	Cf	98	Plutonium	Pu	94
Carbon	C	6	Polonium	Po	84
Cerium	Ce	58	Potassium (Kalium)	K	19
Chlorine	Cl	17	Praseodymium	Pr	59
Chromium	Cr	24	Promethium	Pm	61
Cobalt	Co	27	Protactinium	Pa	91
Copper (Cuprum)	Cu	29	Radium	Ra	88
Curium	Cm	96	Radon	Rn	86
Dysprosium	Dy	66	Rhenium	Re	75
Einsteinium	Es	99	Rhodium	Rh	45
Erbium	Er	68	Rubidium	Rb	37
Europium	Eu	63	Ruthenium	Ru	44
Fermium	Fm	100	Samarium	Sm	62
Fluorine	F	9	Scandium	Sc	21
Francium	Fr	87	Selenium	Se	34
Gadolinium	Gd	64	Silicon	Si	14
Gallium	Ga	31	Silver (Argentum)	Ag	47
Germanium	Ge	32	Sodium (Natrium)	Na	11
Gold (Aurum)	Au	79	Strontium	Sr	38
Hafnium	Hf	72	Sulfur (Theion)[†]	S	16
Helium	He	2	Tantalum	Ta	73
Holmium	Ho	67	Technetium	Tc	43
Hydrogen*	H	1	Tellurium	Te	52
Indium	In	49	Terbium	Tb	65
Iodine	I	53	Thallium	Tl	81
Iridium	Ir	77	Thorium	Th	90
Iron (Ferrum)	Fe	26	Thulium	Tm	69
Krypton	Kr	36	Tin (Stannum)	Sn	50
Lanthanum	La	57	Titanium	Ti	22
Lawrencium (Unniltrium)	Lr	103	Tungsten (Wolfram)	W	74
Lead (Plumbum)	Pb	82	Unnilennium	Une	109
Lithium	Li	3	Unnilhexium	Unh	106
Lutetium	Lu	71	Unniloctium	Uno	108
Magnesium	Mg	12	Unnilpentium	Unp	105

TABLE I

Table I (*Continued*)

Name	Symbol	Atomic number	Name	Symbol	Atomic number
Unnilquadium	Unq	104	Ytterbium	Yb	70
Unnilseptium	Uns	107	Yttrium	Y	39
Uranium	U	92	Zinc	Zn	30
Vanadium	V	23	Zirconium	Zr	40
Xenon	Xe	54			

*The hydrogen isotopes ^2H and ^3H are named deuterium and tritium, respectively, for which the symbols D and T may be used although ^2H and ^3H are preferred (see Section I-3.5.2 and Note 8d).
†This Greek name provides the root 'thi' for sulfur.

TABLE II

Table II Names of atoms [elements] of atomic numbers greater than 100

Atomic number	Name	Symbol
101	Mendelevium (Unnilunium)	Md*
102	Nobelium (Unnilbium)	No*
103	Lawrencium (Unniltrium)	Lr*
104	Unnilquadium	Unq
105	Unnilpentium	Unp
106	Unnilhexium	Unh
107	Unnilseptium	Uns
108	Unniloctium	Uno
109	Unnilennium	Une
110	Ununnilium	Uun
111	Unununium	Uuu
112	Ununbium	Uub
113	Ununtrium	Uut
114	Ununquadium	Uuq
115	Ununpentium	Uup
116	Ununhexium	Uuh
117	Ununseptium	Uus
118	Ununoctium	Uuo
119	Ununennium	Uue
120	Unbinilium	Ubn
121	Unbiunium	Ubu
130	Untrinilium	Utn
140	Unquadnilium	Uqn
150	Unpentnilium	Upn
160	Unhexnilium	Uhn
170	Unseptnilium	Usn
180	Unoctnilium	Uon
190	Unennilium	Uen
200	Binilnilium	Bnn
201	Binilunium	Bnu
202	Binilbium	Bnb
300	Trinilnilium	Tnn
400	Quadnilnilium	Qnn
500	Pentnilnilium	Pnn
900	Ennilnilium	Enn

*To correspond to the systematic names, the systematic symbols would be Unu, Unb and Unt, respectively, but these are not IUPAC approved.

TABLE III

Table III Numerical prefixes

1	mono	19	nonadeca
2	di (bis)	20	icosa
3	tri (tris)	21	henicosa
4	tetra (tetrakis)	22	docosa
5	penta (pentakis)	23	tricosa
6	hexa (hexakis)	30	triaconta
7	hepta (heptakis)	31	hentriaconta
8	octa (octakis)	35	pentatriaconta
9	nona (nonakis)	40	tetraconta
10	deca (decakis), etc.	48	octatetraconta
11	undeca	50	pentaconta
12	dodeca	52	dopentaconta
13	trideca	60	hexaconta
14	tetradeca	70	heptaconta
15	pentadeca	80	octaconta
16	hexadeca	90	nonaconta
17	heptadeca	100	hecta
18	octadeca		

TABLE IV

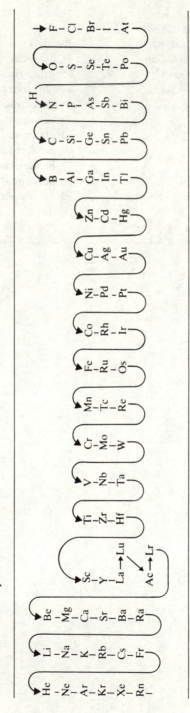

Table IV Element sequence

TABLE V

Table V Structural prefixes used in inorganic nomenclature

These affixes are italicized and separated from the rest of the name by hyphens.

antiprismo	eight atoms bound into a rectangular antiprism
arachno	a boron structure intermediate between *nido*- and *hypho*- in degree of openness
asym	asymmetrical
catena	a chain structure; often used to designate linear polymeric substances
cis	two groups occupying adjacent positions, not now generally recommended for precise nomenclature purposes
closo	a cage or closed structure, especially a boron skeleton that is a polyhedron having all faces triangular
cyclo	a ring structure. Here *cyclo* is used as a modifier indicating structure and hence is italicized. In organic nomenclature, 'cyclo' is considered to be part of the parent name since it changes the molecular formula and therefore is not italicized
dodecahedro	eight atoms bound into a dodecahedron with triangular faces
fac	three groups occupying the corners of the same face of an octahedron, not now generally recommended for precise nomenclature purposes
hexahedro	eight atoms bound into a hexahedron (e.g., cube)
hexaprismo	twelve atoms bound into a hexagonal prism
hypho	an open structure, especially a boron skeleton, more closed than a *klado*-structure, but more open than an *arachno*-structure
icosahedro	twelve atoms bound into a triangular icosahedron
klado	a very open polyboron structure
mer	meridional; three groups occupying vertices of an octahedron in such a relationship that one is *cis* to the two others which are themselves *trans*, not now recommended for precise nomenclature purposes
nido	a nest-like structure, especially a boron skeleton that is almost closed
octahedro	six atoms bound into an octahedron
pentaprismo	ten atoms bound into a pentagonal prism
quadro	four atoms bound into a quadrangle (e.g., square)
sym	symmetrical
tetrahedro	four atoms bound into a tetrahedron
trans	two groups directly across a central atom from each other, i.e., in the polar positions on a sphere, not now generally recommended for precise nomenclature purposes
triangulo	three atoms bound into a triangle
triprismo	six atoms bound into a triangular prism
μ (*mu*)	signifies that a group so designated bridges two or more centres of coordination
λ (*lambda*)	signifies, with its superscript, the bonding number, i.e., the sum of the number of skeletal bonds and the number of hydrogen atoms associated with an atom in a parent compound

TABLE VI

Table VI Seniority list of elements and 'a' terms used in replacement nomenclature, in decreasing order of priority

Element	'a' Term	Element	'a' Term
F	fluora	W	tungsta (wolframa)
Cl	chlora	V	vanada
Br	broma	Nb	nioba
I	ioda	Ta	tantala
At	astata	Ti	titana
O	oxa	Zr	zircona
S	thia	Hf	hafna
Se	selena	Sc	scanda
Te	tellura	Y	yttra
Po	polona	La	lanthana
N	aza	Ce	cera
P	phospha	Pr	praseodyma
As	arsa	Nd	neodyma
Sb	stiba	Pm	prometha
Bi	bisma	Sm	samara
C	carba	Eu	europa
Si	sila	Gd	gadolina
Ge	germa	Tb	terba
Sn	stanna	Dy	dysprosa
Pb	plumba	Ho	holma
B	bora	Er	erba
Al	alumina	Tm	thula
Ga	galla	Yb	ytterba
In	inda	Lu	luteta
Tl	thalla	Ac	actina
Zn	zinca	Th	thora
Cd	cadma	Pa	protactina
Hg	mercura	U	urana
Cu	cupra	Np	neptuna
Ag	argenta	Pu	plutona
Au	aura	Am	america
Ni	nickela	Cm	cura
Pd	pallada	Bk	berkela
Pt	platina	Cf	californa
Co	cobalta	Es	einsteina
Rh	rhoda	Fm	ferma
Ir	irida	Md	mendeleva
Fe	ferra	No	nobela
Ru	ruthena	Lr	lawrenca
Os	osma	Be	berylla
Mn	mangana	Mg	magnesa
Tc	techneta	Ca	calca
Re	rhena	Sr	stronta
Cr	chroma	Ba	bara
Mo	molybda	Ra	rada

Table VII Element substituent group names[a]

Element name	Radical name	Element name	Radical name
Actinium	Actinio	Mercury	Mercurio
Aluminium	Aluminio	Molybdenum	Molybdenio
Americium	Americio	Neodymium	Neodymio
Antimony	Antimonio	Neon	Neonio
Argon	Argonio	Neptunium	Neptunio
Arsenic	Arsenio	Nickel	Nickelio
Astatine	Astatio	Niobium	Niobio
Barium	Bario	Nitrogen	—
Berkelium	Berkelio	Nobelium	Nobelio
Beryllium	Beryllio	Osmium	Osmio
Bismuth	Bismuthio	Oxygen	—
Boron	Borio	Palladium	Palladio
Bromine	Bromio	Phosphorus	Phosphorio
Cadmium	Cadmio	Platinum	Platinio
Caesium	Caesio	Plutonium	Plutonio
Calcium	Calcio	Polonium	Polonio
Californium	Californio	Potassium	Potassio (Kalio)
Carbon	—	Praseodymium	Praseodymio
Cerium	Cerio	Promethium	Promethio
Chlorine	Chlorio	Protactinium	Protactinio
Chromium	Chromio	Radium	Radio
Cobalt	Cobaltio	Radon	Radonio
Copper (Cuprum)	Cuprio	Rhenium	Rhenio
Curium	Curio	Rhodium	Rhodio
Deuterium	Deuterio	Rubidium	Rubidio
Dysprosium	Dysprosio	Ruthenium	Ruthenio
Einsteinium	Einsteinio	Samarium	Samario
Erbium	Erbio	Scandium	Scandio
Europium	Europio	Selenium	Selenio
Fermium	Fermio	Silicon	Silicio
Fluorine	Fluorio	Silver (Argentum)	Argentio
Francium	Francio	Sodium	Sodio (Natrio)
Gadolinium	Gadolinio	Strontium	Strontio
Gallium	Gallio	Sulfur	Sulfurio
Germanium	Germanio	Tantalum	Tantalio
Gold (Aurum)	Aurio	Technetium	Technetio
Hafnium	Hafnio	Tellurium	Tellurio
Helium	Helio	Terbium	Terbio
Holmium	Holmio	Thallium	Thallio
Hydrogen	—	Thorium	Thorio
Indium	Indio	Thulium	Thulio
Iodine	Iodio	Tin (Stannum)	Stannio
Iridium	Iridio	Titanium	Titanio
Iron (Ferrum)	Ferrio	Tritium	Tritio
Krypton	Kryptonio	Tungsten	Tungstenio
Lanthanum	Lanthanio	(Wolfram)	(Wolframio)
Lawrencium	Lawrencio	Uranium	Uranio
Lead (Plumbum)	Plumbio	Vanadium	Vanadio
Lithium	Lithio	Xenon	Xenonio
Lutetium	Lutetio	Ytterbium	Ytterbio
Magnesium	Magnesio	Yttrium	Yttrio
Manganese	Manganio	Zinc	Zincio
Mendelevium	Mendelevio	Zirconium	Zirconio

[a]These names are used in organic substitutive nomenclature for situations in which the substituent group is joined to the parent skeleton by a single element–carbon bond.

TABLE VIII

Table VIII Names of ions and groups*

| Neutral atom or group formula | Name | | | |
	Uncharged (atom, molecule, or radical)	Cation or cationic group	Anion	Ligand
1	2	3	4	5
Ac	actinium	actinium	actinide	
Ag	silver	silver	argentide	
Al	aluminium	aluminium	aluminide	
Am	americium	americium	americide	
Ar	(mono)argon	argon	argonide	
As	(mono)arsenic	arsenic	arsenide	arsenido
AsH_4		AsH_4^+ arsonium		
AsO_3			AsO_3^{3-} arsenite trioxoarsenate(3−) trioxoarsenate(III)	arsenito(3−) trioxoarsenato(3−) trioxoarsenato(III)
AsO_4			AsO_4^{3-} arsenate tetraoxoarsenate(3−) tetraoxoarsenate(v)	arsenato(3−) tetraoxoarsenato(3−) tetraoxoarsenato(v)
AsS_4			AsS_4^{3-} tetrathioarsenate(3−) tetrathioarsenate(v)	tetrathioarsenato(3−) tetrathioarsenato(v)
At	(mono)astatine	astatine	astatide	

248

TABLE VIII

Au	gold	Au^+ gold(1+) gold(I) Au^{3+} gold(3+) gold(III)	auride	
B	(mono)boron boron		boride	borido
BO_2		$(BO_2^-)_n$ metaborate poly[dioxoborate(1−)] poly[dioxoborate(III)]		metaborato
BO_3		BO_3^{3-} borate trioxoborate(3−) trioxoborate(III)		borato trioxoborato(3−) trioxoborato(III)
Ba	barium	barium	baride	
Be	beryllium	beryllium	beryllide	

*The above table contains five columns, the first of which contains the symbol or formula of the neutral atom or group. The second column contains the corresponding name. The third column contains the name corresponding to the symbol or formula when it carries one or more units of positive charge. Inorganic nomenclature allows charges to be represented by the charge number, or to be inferred from an appropriate oxidation number. Both methods are displayed in the third column and in the succeeding columns. Formulae for ions are shown for cases where it is felt that confusion might otherwise arise. The fourth column contains the name of the symbol or formula when it carries one or more units of negative charge. Finally, the fifth column contains the name of the formula or symbol when the species it represents is a ligand (usually assumed to be anionic if it is not neutral).

The symbols (formulae) are listed in alphabetical order according to the principles outlined in Section I-4.6.1.3. Because the terminations -ous and -ic for metal cation names are no longer recommended, these have been excluded, but we have attempted to include all those traditional names which are still allowed. We have not attempted to present names for species of very rare or unlikely occurrence, so that there are gaps in the columns.

Users should note that we name only one specific structure for a given formula. In some cases there may be other structures which we have not named corresponding to that formula.

Table VIII (*Continued*)

Neutral atom or group formula *1*	Uncharged (*atom, molecule, or radical*) *2*	Name — Cation or cationic group *3*	Name — Anion *4*	Name — Ligand *5*
Bi	bismuth	bismuth	bismuthide	bismuthido
Bk	berkelium	berkelium	berkelide	
Br	(mono)bromine	bromine	bromide	bromo
BrO	bromine monoxide	bromosyl	BrO^- oxobromate(1−) oxobromate(i) (not hypobromite)	oxobromato(1−) oxobromato(i)
BrO2	bromine dioxide	bromyl	BrO_2^- dioxobromate(1−) dioxobromate(iii) (not bromite)	dioxobromato(1−) dioxobromato(iii)
BrO3	bromine trioxide	perbromyl	BrO_3^- trioxobromate(1−) trioxobromate(v) (not bromate)	trioxobromato(1−) trioxobromato(v)
BrO4	bromine tetraoxide		BrO_4^- tetraoxobromate(1−) tetraoxobromate(vii) (not perbromate)	tetraoxobromato(1−) tetraoxobromato(vii)
Br3	tribromine		tribromide(1−)	tribromo(1−)
C	(mono)carbon	carbon	carbide	carbido
CN			CN^- cyanide	cyano

TABLE VIII

Formula				
CO	carbon monoxide	carbonyl		carbonyl carbon monoxide
CO_3			$CO_3{}^{2-}$ carbonate trioxocarbonate(2−) trioxocarbonate(iv)	carbonato trioxocarbonato(2−) trioxocarbonato(iv)
CS	carbon monosulfide	thiocarbonyl		thiocarbonyl carbon monosulfide
CS_3			$CS_3{}^{2-}$ trithiocarbonate(2−) trithiocarbonate(iv)	trithiocarbonato(2−) trithiocarbonato(iv)
C_2	dicarbon		$C_2{}^{2-}$ acetylide dicarbide(2−)	dicarbido
Ca	calcium	calcium	calcide	
Cd	cadmium	cadmium	cadmide	
Ce	cerium	cerium	ceride	
Cf	californium	californium	californide	
Cl	(mono)chlorine	chlorine	chloride	chloro
ClF_4	chlorine tetrafluoride	$ClF_4{}^+$ tetrafluorochlorine(1+) tetrafluorochlorine(v)	$ClF_4{}^-$ tetrafluorochlorate(1−) tetrafluorochlorate(iii)	tetrafluorochlorato(1−) tetrafluorochlorato(iii)
ClO	chlorine monoxide	chlorosyl	ClO^- hypochlorite oxochlorate(1−) oxochlorate(i)	hypochlorito oxochlorato(1−) oxochlorato(i)
ClO_2	chlorine dioxide	chloryl	$ClO_2{}^-$ chlorite dioxochlorate(1−) dioxochlorate(iii)	chlorito dioxochlorato(1−) dioxochlorato(iii)

Table VIII (*Continued*)

Neutral atom or group formula 1	Uncharged (atom, molecule, or radical) 2	Name		
		Cation or cationic group 3	*Anion* 4	*Ligand* 5
ClO_3	chlorine trioxide	perchloryl	ClO_3^- chlorate trioxochlorate(1−) trioxochlorate(v)	chlorato trioxochlorato(1−) trioxochlorato(v)
ClO_4	chlorine tetraoxide		ClO_4^- perchlorate tetraoxochlorate(1−) tetraoxochlorate(vii)	perchlorato tetraoxochlorato(1−) tetraoxochlorato(vii)
Cm	curium	curium	curide	
Co	cobalt	Co^{2+} cobalt(2+) cobalt(ii) Co^{3+} cobalt(3+) cobalt(iii)	cobaltide	
Cr	chromium	Cr^{2+} chromium(2+) chromium(ii) Cr^{3+} chromium(3+) chromium(iii)	chromide	
CrO_2	chromium dioxide	chromyl		

TABLE VIII

CrO_4			CrO_4^{2-} chromate tetraoxochromate(2−) tetraoxochromate(vi)	chromato tetraoxochromato(2−) tetraoxochromato(vi)
Cr_2O_7			$Cr_2O_7^{2-}$ dichromate(2−) μ-oxo-hexaoxo-dichromate(2−) μ-oxo-hexaoxo-dichromate(vi)	dichromato(2−) μ-oxo-hexaoxo-dichromato(2−) μ-oxo-hexaoxo-dichromato(vi)
Cs	caesium	caesium	caeside	
Cu	copper	Cu^+ copper(1+) copper(i) Cu^{2+} copper(2+) copper(ii)	cupride	
Dy	dysprosium	dysprosium	dysproside	
Er	erbium	erbium	erbide	
Es	einsteinium	einsteinium	einsteinide	
Eu	europium	europium	europide	
F	(mono)fluorine	fluorine	fluoride	fluoro
Fe	iron	Fe^{2+} iron(2+) iron(ii) Fe^{3+} iron(3+) iron(iii)	ferride	
Fm	fermium	fermium	fermide	
Fr	francium	francium	francide	

Table VIII (*Continued*)

Neutral atom or group formula *1*	Uncharged (atom, molecule, or radical) *2*	Cation or cationic group *3*	Name	
			Anion *4*	Ligand *5*
Ga	gallium	gallium	gallide	
Gd	gadolinium	gadolinium	gadolinide	
Ge	germanium	Ge^{2+} germanium(2+) germanium(II) Ge^{4+} germanium(4+) germanium(IV)	germide	
H	(mono)hydrogen	hydrogen	hydride	hydrido hydro (in boron compounds)
HCO_3			HCO_3^- hydrogencarbonate(1−) hydrogentrioxo-carbonate(1−) hydrogentrioxo-carbonate(IV)	hydrogencarbonato(1−) hydrogentrioxo-carbonato(1−) hydrogentrioxo-carbonato(IV)
HO	HO hydroxyl	HO^+ hydroxylium	OH^- hydroxide	hydroxido hydroxo
HO_2	hydrogen dioxide	hydrogenperoxyl perhydroxyl hydroperoxyl	HO_2^- hydrogenperoxide(1−) hydrogendioxide(1−)	hydrogenperoxo

TABLE VIII

HPO_4^{2-}	hydrogenphosphate(2−) hydrogentetraoxophosphate(2−) hydrogentetraoxophosphate(v)	hydrogenphosphato(2−) hydrogentetraoxophosphato(2−) hydrogentetraoxophosphato(v)
HS^-	hydrogensulfide(1−)	hydrogensulfido(1−) sulfanido
HSO_3^-	hydrogensulfite(1−) hydrogentrioxosulfate(1−) hydrogentrioxosulfate(iv)	hydrogensulfito(1−) hydrogentrioxosulfato(1−) hydrogentrioxosulfato(iv)
HSO_4^-	hydrogensulfate(1−) hydrogentetraoxosulfate(1−) hydrogentetraoxosulfate(vi)	hydrogensulfato(1−) hydrogentetraoxosulfato(1−) hydrogentetraoxosulfato(vi)
HSe^-	hydrogenselenide(1−)	hydrogenselenido selanido
$HSeO_3^-$	hydrogentrioxoselenate(1−) hydrogentrioxoselenate(iv)	
HTe^-	hydrogentelluride(1−)	hydrogentellurido tellanido

Table VIII (*Continued*)

TABLE VIII

Neutral atom or group formula 1	Uncharged (atom, molecule, or radical) 2	Name		
		Cation or cationic group 3	Anion 4	Ligand 5
H_2Br		H_2Br^+ bromonium		
H_2Cl		H_2Cl^+ chloronium		
H_2F		H_2F^+ fluoronium		
H_2I		H_2I^+ iodonium		
H_2NO, see NHOH				
H_2O	oxidane water			aqua oxidane
H_2O_2P, see PH_2O_2				
$H_2O_5P_2$, see $P_2H_2O_5$				
H_2PO_4			$H_2PO_4^-$ dihydrogenphosphate(1−) dihydrogentetraoxo-phosphate(1−) dihydrogentetraoxophosphate(v)	dihydrogenphosphato(1−) dihydrogentetraoxophosphato(1−) dihydrogentetraoxophosphato(v)
H_3O	trihydrogen oxide	H_3O^+ oxonium		

TABLE VIII

H₃S	trihydrogen sulfide	H₃S⁺ sulfonium		
H₃Se	trihydrogen selenide	H₃Se⁺ selenonium		
H₃Te	trihydrogen telluride	H₃Te⁺ telluronium		
He	(mono)helium	helium	helide	
Hf	hafnium	hafnium	hafnide	
Hg	mercury	Hg²⁺ mercury(2+) mercury(II); Hg₂²⁺ dimercury(2+) dimercury(I)	mercuride	
Ho	holmium	holmium	holmide	
I	(mono)iodine	iodine	iodide	iodo
IF₄	iodine tetrafluoride	IF₄⁺ tetrafluoroiodine(1+) tetrafluoroiodine(v)	IF₄⁻ tetrafluoroiodate(1−) tetrafluoroiodate(III)	tetrafluoroiodato(1−) tetrafluoroiodato(III)
IO	iodine oxide	iodosyl	IO⁻ oxoiodate(1−) oxoiodate(I) (not hypoiodite)	oxoiodato(1−) oxoiodato(I)
IO₂	iodine dioxide	iodyl	IO₂⁻ dioxoiodate(1−) dioxoiodate(III) (not iodite)	dioxoiodato(1−) dioxoiodato(III)
IO₃	iodine trioxide	periodyl	IO₃⁻ iodate trioxoiodate(1−) trioxoiodate(v)	iodato trioxoiodato(1−) trioxoiodato(v)

Table VIII (*Continued*)

| Neutral atom or group formula | Uncharged (atom, molecule, or radical) | Name | | |
		Cation or cationic group	Anion	Ligand
1	2	3	4	5
IO_4	iodine tetraoxide		IO_4^- periodate, tetraoxoiodate(1−), tetraoxoiodate(VII)	periodato, tetraoxoiodato(1−), tetraoxoiodato(VII)
IO_6			IO_6^{5-} hexaoxoiodate(5−), hexaoxoiodate(VII)	hexaoxoiodato(5−), hexaoxoiodato(VII)
I_3	triiodine		triiodide(1−)	triiodo(1−)
In	indium	indium	indide	
Ir	iridium	iridium	iridide	
K	potassium	potassium	kalide	
Kr	(mono)krypton	krypton	kryptonide	
La	lanthanum	lanthanum	lanthanide	
Li	lithium	lithium	lithide	
Lr	lawrencium	lawrencium	lawrencide	
Lu	lutetium	lutetium	lutetide	
Md	mendelevium	mendelevium	mendelevide	
Mg	magnesium	magnesium	magneside	

TABLE VIII

TABLE VIII

Mn	manganese	Mn^{2+} manganese(2+) manganese(II) Mn^{3+} manganese(3+) manganese(III)	manganide	
MnO_4		MnO_4^- permanganate tetraoxomanganate(1−) tetraoxomanganate(VII) MnO_4^{2-} manganate tetraoxomanganate(2−) tetraoxomanganate(VI)	permanganato tetraoxomanganato(1−) tetraoxomanganato(VII) manganato tetraoxomanganato(2−) tetraoxomanganato(VI)	
Mo	molybdenum	molybdenum	molybdenide	
N	(mono)nitrogen	nitrogen	nitride	nitrido
NCO, see OCN				
NF_4		NF_4^+ tetrafluoroammonium tetrafluoronitrogen(1+) tetrafluoronitrogen(V)		
NH			NH^{2-} imide azanediide azanide(2−)	imido azanediido
NHOH			$NHOH^-$ hydroxyamide	hydroxyamido
NH_2			NH_2^- amide azanide	amido azanido

TABLE VIII

Table VIII (*Continued*)

Neutral atom or group formula	Uncharged (atom, molecule, or radical)	Name		
		Cation or cationic group	*Anion*	*Ligand*
1	*2*	*3*	*4*	*5*
NH_3	azane ammonia	NH_3^+ ammoniumyl azaniumyl		ammine azane
NH_4		NH_4^+ ammonium azanium		
NO	nitrogen monoxide	nitrosyl	NO^- oxonitrate(1−) oxonitrate(I)	nitrosyl nitrogen monoxide
NO_2	nitrogen dioxide	nitryl nitroyl	NO_2^- dioxonitrate(1−) dioxonitrate(III) NO_2^{2-} dioxonitrate(2−) dioxonitrate(II) (not nitroxylate)	nitro nitrito-*O* nitrito-*N* dioxonitrato(1−) dioxonitrato(III) dioxonitrato(2−) dioxonitrato(II)

TABLE VIII

	Neutral	Cation	Anion	Ligand
NO₃	nitrogen trioxide		NO_3^- nitrate, trioxonitrate(1−), trioxonitrate(v)	nitrato, trioxonitrato(1−), trioxonitrato(v)
NO₄			$NO(O_2)^-$ oxoperoxonitrate(1−), oxoperoxonitrate(iii) (not peroxonitrite); $NO_2(O_2)^-$ dioxoperoxonitrate(1−), dioxoperoxonitrate(v) (not peroxonitrate)	
NS	nitrogen monosulfide	thionitrosyl	thionitrosyl	thionitrosyl, nitrogen monosulfide
N₂H		N_2H^+ diazynium	N_2H^- diazenide; N_2H^{3-} diazanetriide, diazanide(3−), hydrazinetriide, hydrazinide(3−), hydrazide(3−)	diazenido; diazanetriido, hydrazido(3−)
N₂H₂	diazene, diimide	$N_2H_2^{2+}$ diazynediium, diazynium(2+)	$N_2H_2^{2-}$ diazanediide, hydrazide(2−), diazanide(2−), hydrazinediide	diazanediido, hydrazido(2−); N_2H_2 diazene, diimide

TABLE VIII

Table VIII (*Continued*)

Neutral atom or group formula *1*	Uncharged (atom, molecule, or radical) *2*	Cation or cationic group *3*	Name	
			Anion *4*	Ligand *5*
NHNH$_2$		N$_2$H$_3^+$ diazenium	N$_2$H$_3^-$ hydrazide diazanide hydrazinide	hydrazido diazanido
N$_2$H$_4$	diazane hydrazine	N$_2$H$_4^{2+}$ diazenediium diazenium(2+)		hydrazine diazane
N$_2$H$_5$		N$_2$H$_5^+$ hydrazinium(1+) diazanium		hydrazinium
N$_2$H$_6$		N$_2$H$_6^{2+}$ hydrazinium(2+) diazanediium diazanium(2+) hydrazinediium		
N$_2$O$_2$	dinitrogen dioxide		N$_2$O$_2^{2-}$ dioxodinitrate(N—N)(2−) dioxodinitrate(N—N)(1t) (not hyponitrite)	dioxodinitrato(N—N)(2−) dioxodinitrato(N—N)(1t)
N$_3$	trinitrogen	trinitrogen	azide trinitride(1−)	azido trinitrido(1−)
Na	sodium	sodium	natride	
Nb	niobium	niobium	niobide	

TABLE VIII

Formula				
Nd	neodymium	neodymium	neodymide	
Ne	(mono)neon	neon	neonide	
Ni	nickel	Ni^{2+}	nickelide	
		nickel(2+)		
		nickel(II)		
		Ni^{3+}		
		nickel(3+)		
		nickel(III)		
No	nobelium	nobelium	nobelide	
Np	neptunium	neptunium	neptunide	
NpO_2	neptunium dioxide	neptunyl		
O	(mono)oxygen	oxygen	oxide	oxo
				oxido
OCN			cyanate	cyanato
				cyanato-*O*
				cyanato-*N*
			nitridooxocarbonate(1−)	nitridooxocarbonato(1−)
			nitridooxocarbonate(IV)	nitridooxocarbonato(IV)
OH, see HO				
ONC			fulminate	fulminato
			carbidooxonitrate(1−)	carbidooxonitrato(1−)
			carbidooxonitrate(V)	carbidooxonitrato(V)
O_2	dioxygen	O_2^+	O_2^{2-}	peroxo
		dioxygen(1+)	peroxide	dioxido(2−)
			dioxide(2−)	
			O_2^-	hyperoxo
			hyperoxide	superoxido
			superoxide	dioxido(1−)
			dioxide(1−)	O_2
				dioxygen

263

TABLE VIII

Table VIII (*Continued*)

Neutral atom or group formula 1	Uncharged (atom, molecule, or radical) 2	Cation or cationic group 3	Name	
			Anion 4	Ligand 5
O_3	trioxygen ozone		O_3^- ozonide trioxide(1−)	ozonido trioxido(1−) O_3 trioxygen
Os	osmium	osmium	osmide	
P	(mono)phosphorus	phosphorus	P^{3-} phosphide	phosphido
PCl_4	phosphorus tetrachloride	PCl_4^+ tetrachlorophosphonium tetrachlorophosphonium(v) tetrachlorophosphorus(1+) tetrachlorophosphorus(v) tetrachlorophosphanium(1+)	PCl_4^- tetrachlorophosphate(1−) tetrachlorophosphate(III)	tetrachlorophosphato(1−) tetrachlorophosphato(III)
PHO_3			PHO_3^{2-} phosphonate hydridotrioxophosphate(2−)	phosphonato(2−) hydridotrioxophosphato(2−)
PH_2O_2			$PH_2O_2^-$ phosphinate dihydridodioxophosphate(1−)	phosphinato dihydridodioxophosphato(1−)
PH_4		PH_4^+ phosphonium		
PO	phosphorus monoxide	phosphoryl		

TABLE VIII

Formula			
PO_2	phosphorus dioxide	PO_2^- dioxophosphate(1−) dioxophosphate(III)	
PO_3		PO_3^{3-} phosphite trioxophosphate(3−) trioxophosphate(III) $(PO_3^-)_n$ metaphosphate poly[trioxophosphate(1−)] poly[trioxophosphate(v)]	phosphito(3−) trioxophosphato(3−) trioxophosphato(III)
PO_4		PO_4^{3-} phosphate orthophosphate tetraoxophosphate(3−) tetraoxophosphate(v)	phosphato(3−) orthophosphato tetraoxophosphato(3−) tetraoxophosphato(v)
PO_5		PO_5^{3-} trioxoperoxophosphate(3−) trioxoperoxophosphate(v) (not peroxomonophosphate)	
PS	phosphorus sulfide	thiophosphoryl	
PS_4		PS_4^{3-} tetrathiophosphate(3−) tetrathiophosphate(v)	
$P_2H_2O_5$		$P_2H_2O_5^{2-}$ diphosphonate	
P_2O_6 $(O_3P{-}PO_3)$		$P_2O_6^{4-}$ hypophosphate hexaoxodiphosph- ate $(P{-}P)(4-)$ hexaoxodiphosph- ate $(P{-}P)$(IV)	

Table VIII (*Continued*)

			Name	
Neutral atom or group formula	*Uncharged (atom, molecule, or radical)*	*Cation or cationic group*	*Anion*	*Ligand*
1	*2*	*3*	*4*	*5*
P_2O_7	diphosphorus heptaoxide		$P_2O_7{}^{4-}$ diphosphate(4−) μ-oxo-hexaoxodiphosphate(4−) μ-oxo-hexaoxodiphosphate(v)	diphosphato(4−) μ-oxo-hexaoxodiphosphato(4−) μ-oxo-hexaoxodiphosphato(v)
P_2O_8			$P_2O_8{}^{4-}$ hexaoxo-μ-peroxo-diphosphate(4−) hexaoxo-μ-peroxo-diphosphate(v) (not peroxodiphosphate)	
Pa	protactinium	protactinium	protactinide	
Pb	lead	Pb^{2+} lead(2+) lead(ii) Pb^{4+} lead(4+) lead(iv)	plumbide	
Pd	palladium	Pd^{2+} palladium(2+) palladium(ii) Pd^{4+} palladium(4+) palladium(iv)	palladide	

TABLE VIII

Pm	promethium	promethium	promethide	
Po	polonium	polonium	polonide	
Pr	praseodymium	praseodymium	praseodymide	
Pt	platinum	platinum	platinide	
		Pt^{2+}		
		platinum(2+)		
		platinum(ii)		
		Pt^{4+}		
		platinum(4+)		
		platinum(iv)		
Pu	plutonium	plutonium	plutonide	
PuO_2	plutonium dioxide	plutonyl		
Ra	radium	radium	radide	
Rb	rubidium	rubidium	rubidide	
Re	rhenium	rhenium	rhenide	
ReO_4			ReO_4^-	
			tetraoxorhenate(1−)	tetraoxorhenato(1−)
			tetraoxorhenate(vii)	tetraoxorhenato(vii)
			(not perrhenate)	
			ReO_4^{2-}	
			tetraoxorhenate(2−)	tetraoxorhenato(2−)
			tetraoxorhenate(vi)	tetraoxorhenato(vi)
			(not rhenate)	
Rh	rhodium	rhodium	rhodide	
Rn	(mono)radon	radon	radonide	
Ru	ruthenium	ruthenium	ruthenide	
S	(mono)sulfur	sulfur	sulfide	sulfido
				thio
SCN			thiocyanate	thiocyanato-N
				thiocyanato-S
			nitridothiocarbonate(1−)	nitridothiocarbonato(1−)
			nitridothiocarbonate(iv)	nitridothiocarbonato(iv)

Table VIII (*Continued*)

Neutral atom or group formula	*Uncharged (atom, molecule, or radical)*	*Cation or cationic group*	*Name*	
			Anion	*Ligand*
1	*2*	*3*	*4*	*5*
SO	sulfur monoxide	sulfinyl thionyl		sulfur monoxide
SO_2	sulfur dioxide	sulfonyl sulfuryl	$SO_2{}^{2-}$ dioxosulfate(2−) dioxosulfate(II) (not sulfoxylate)	dioxosulfato(2−) dioxosulfato(II) SO_2 sulfur dioxide
SO_3	sulfur trioxide		$SO_3{}^{2-}$ sulfite trioxosulfate(2−) trioxosulfate(IV)	sulfito trioxosulfato(2−) trioxosulfato(IV)
SO_4	sulfur tetraoxide		$SO_4{}^{2-}$ sulfate tetraoxosulfate(2−) tetraoxosulfate(VI)	sulfato tetraoxosulfato(2−) tetraoxosulfato(VI)
SO_5			$SO_5{}^{2-}$ trioxoperoxosulfate(2−) trioxoperoxosulfate(VI) (not peroxomonosulfate)	
S_2	disulfur		$S_2{}^{2-}$ disulfide(2−)	disulfido(2−)

TABLE VIII

Formula		Anion names	Ligand names
S_2O_2		$S_2O_2^{2-}$	
		dioxothiosulfate(2−)	dioxothiosulfato(2−)
		dioxothiosulfate(IV)	dioxothiosulfato(IV)
		(not thiosulfite)	
S_2O_3	disulfur trioxide	$S_2O_3^{2-}$	
		thiosulfate	thiosulfato
		trioxothiosulfate(2−)	trioxothiosulfato(2−)
		trioxothiosulfate(VI)	trioxothiosulfato(VI)
S_2O_4		$S_2O_4^{2-}$	
		dithionite	dithionito
		tetraoxodisulfate(S—S)(2−)	tetraoxodisulfato(S—S)(2−)
		tetraoxodisulfate(S—S)(III)	tetraoxodisulfato(S—S)(III)
S_2O_5	disulfur pentaoxide	$S_2O_5^{2-}$	
		μ-oxo-tetraoxodisulfate(2−)	
		μ-oxo-tetraoxodisulfate(IV)	
		(not disulfite)	
S_2O_6		$S_2O_6^{2-}$	
		dithionate	dithionato
		hexaoxodisulfate(S—S)(2−)	hexaoxodisulfato(S—S)(2−)
		hexaoxodisulfate(S—S)(V)	hexaoxodisulfato(S—S)(V)
S_2O_7	disulfuryl	$S_2O_7^{2-}$	
		disulfate(2−)	disulfato(2−)
		μ-oxo-hexaoxodisulfate(2−)	μ-oxo-hexaoxodisulfato(2−)
		μ-oxo-hexaoxodisulfate(VI)	μ-oxo-hexaoxodisulfato(VI)
S_2O_8		$S_2O_8^{2-}$	
		μ-peroxo-hexaoxo-disulfate(2−)	
		μ-peroxo-hexaoxo-disulfate(VI)	
		(not peroxodisulfate)	

Table VIII (*Continued*)

Neutral atom or group formula 1	Uncharged (atom, molecule, or radical) 2	Name Cation or cationic group 3	Anion 4	Ligand 5
S_4	tetrasulfur	S_4^{2+} tetrasulfur(2+)	tetrasulfide(2−)	tetrasulfido(2−)
Sb	(mono)antimony	antimony	antimonide	antimonido
SbH_4		SbH_4^+ stibonium		
Sc	scandium	scandium	scandide	
Se	(mono)selenium	selenium	selenide	selenido
SeCN			selenocyanate nitridoselenido-carbonate(1−) nitridoselenido-carbonate(IV)	selenocyanato nitridoselenido-carbonato(1−) nitridoselenido-carbonato(IV)
SeO	selenium monoxide	seleninyl		selenium monoxide
SeO_2	selenium dioxide	selenonyl	SeO_2^{2-} dioxoselenate(2−) dioxoselenate(II)	dioxoselenato(2−) dioxoselenato(II)
SeO_3	selenium trioxide		SeO_3^{2-} trioxoselenate(2−) trioxoselenate(IV) (not selenite)	trioxoselenato(2−) trioxoselenato(IV)

TABLE VIII

Formula				
SeO_4			SeO_4^{2-} tetraoxoselenate(2−) tetraoxoselenate(VI) (not selenate)	tetraoxoselenato(2−) tetraoxoselenato(VI)
Si	(mono)silicon	silicon	silicide	silicido
SiO_3			$(SiO_3^{2-})_n$ metasilicate poly[trioxosilicate(2−)] poly[trioxosilicate(IV)]	
SiO_4			SiO_4^{4-} orthosilicate tetraoxosilicate(4−) tetraoxosilicate(IV)	
Si_2O_7			$Si_2O_7^{6-}$ μ-oxo-hexaoxo-disilicate(6−) μ-oxo-hexaoxo-disilicate(IV)	
Sm	samarium	samarium	samaride	
Sn	tin	Sn^{2+} tin(2+) tin(II) Sn^{4+} tin(4+) tin(IV)	stannide	
Sr	strontium	strontium	strontide	
Ta	tantalum	tantalum	tantalide	
Tb	terbium	terbium	terbide	
Tc	technetium	technetium	technetide	

Table VIII (*Continued*)

Neutral atom or group formula *1*	Uncharged (atom, molecule, or radical) *2*	Name		
		Cation or cationic group *3*	Anion *4*	Ligand *5*
TcO_4			TcO_4^- tetraoxotechnetate(1−) tetraoxotechnetate(vii) (not pertechnetate) TcO_4^{2-} tetraoxotechnetate(2−) tetraoxotechnetate(vi) (not technetate)	
Te	(mono)tellurium	tellurium	telluride	tellurido
TeO_3			TeO_3^{2-} trioxotellurate(2−) trioxotellurate(iv)	
TeO_4			TeO_4^{2-} tetraoxotellurate(2−) tetraoxotellurate(vi)	
TeO_6			TeO_6^{6-} hexaoxotellurate(6−) hexaoxotellurate(vi) (not orthotellurate)	hexaoxotellurato(6−) hexaoxotellurato(vi)
Th	thorium	thorium	thoride	
Ti	titanium	titanium	titanide	
TiO	titanium monoxide	oxotitanium(iv)		
Tl	thallium	thallium	thallide	

TABLE VIII

Tm	thulium	thulium	thulide
U	uranium	uranium	uranide
UO$_2$	uranium dioxide	UO$_2^+$	
		uranyl(1+)	
		uranyl(v)	
		dioxouranium(1+)	
		dioxouranium(v)	
		UO$_2^{2+}$	
		uranyl(2+)	
		uranyl(vi)	
		dioxouranium(2+)	
		dioxouranium(vi)	
V	vanadium	vanadium	vanadide
VO	vanadium monoxide	oxovanadium(iv)	
W	tungsten	tungsten	tungstide
Xe	(mono)xenon	xenon	xenonide
Y	yttrium	yttrium	yttride
Yb	ytterbium	ytterbium	ytterbide
Zn	zinc	zinc	zincide
Zr	zirconium	zirconium	zirconide
ZrO	zirconium monoxide	oxozirconium(iv)	

TABLE IX

Table IX A selection of affixes used in inorganic and organic nomenclature

1. Simple affixes

-a	Termination vowel for skeletal replacement nomenclature
	In Hantzsch–Widman nomenclature: -oxa, -aza
	In boron nomenclature for heteroatoms: -carba, -thia
-ane	Termination for names of neutral saturated hydrides of boron, and of elements of Groups 14, 15, and 16: diphosphane
	Hantzsch–Widman termination for names of all saturated *n*-membered carbon rings
-ate	General suffix for many polyatomic anions in inorganic nomenclature (including coordination nomenclature) and in organic nomenclature: nitrate, acetate, hexacyanoferrate. There are some anions which are exceptions, having names which end in -ite or -ide
-ene	Termination for names of unsaturated acyclic and cyclic hydrocarbons: pentene, cyclohexene
	Termination for names of unsaturated homogeneous and heterogeneous chain and ring compounds: triazene
-enic	Termination for suffix indicating an acid of lower oxidation state than the -inic acid: phosphenic acid
-eno	Termination for prefix indicating an acid of lower oxidation state than the -ino: phospheno-
-ic	Termination for names of many acids both inorganic and organic: sulfuric acid, benzoic acid
-ide	Termination for names of certain monoatomic anions: chloride, sulfide
	Termination for names of the more electronegative constituent in binary type names: disulfur dichloride
	In names of homopolyatomic anions: triiodide
	In certain heteropolyatomic anions: cyanide
	In those anions formed by removal of one or more hydrogen ions from a molecular hydride, or from an organic derivative thereof: hydrazinide
	Termination for a systematic name of an organic anion: methanide
-ine	Termination for trivial names of certain hydrides such as N_2H_4 and PH_3: hydrazine, phosphine: a name termination in the Hantzsch–Widman System
	Termination for many trivial names for heterocyclic compounds: -iridine, -etidine, -olidine
-inico	Termination for prefixes indicating an oxoacid name ending with -inic: arsinico
-ino	Termination for prefixes indicating an oxoacid: sulfino-
-inoyl	Termination for prefixes indicating a radical of the type $H_2X(O)$: phosphinoyl
-io	General termination for radicals and substituent groups of all kinds containing a metal centre from which the linkage is made; such groups can be either coordination entities or organometallic entities: cuprio-, methylmercurio-, tetra-carbonylcobaltio-
	Termination for prefixes indicating a cationic centre in a structure: -onium becomes -onio, as in ammonio-; -inium becomes -inio, as in pyridinio-
-ite	Termination for anions (esters and salts) of certain oxoacids having the -ous ending in the acid name: sulfite. This usage is discouraged unless listed for present use

TABLE IX

Table IX (*Continued*)

-ium	Termination of names for many elements, and preferred termination for the name of any new element
	Termination for many electropositive constituents in binary type names, either inorganic or organic, either systematic or trivial
	Termination indicating the addition of one hydrogen ion (or a positive alkyl group) to a molecular hydride or its substitution product: ammonium, phosphanium
	Termination for names of cations formed from metallocenes: ferrocenium
-o	Termination indicating a negatively charged ligand: bromo-. Usually it appears as -ido, -ito, -ato
	Termination for the names of many inorganic and organic radicals: chloro-, piperidino-
	Termination for infixes used in infix nomenclature to indicate replacement of oxygen atoms and/or hydroxyl groups: thio-, nitrido-
-ocene	Suffix for the trivial names of bis(cyclopentadienyl)metals and their derivatives: ferrocene
-onate	Termination for name of an anion deriving from an -onic acid
-onic	Termination for names of acids of the type RSO_2OH or $RPO(OH)_2$: sulfonic acid, phosphonic acid (R = H, alkyl, or aryl)
-onite	Termination for names of anions or esters deriving from $RP(OH)_2$ or similar compounds (R = H or alkyl)
-ono	Termination for prefixes indicating an oxoacid of connectivity 4 with one H or alkyl group attached to the central atom: phosphono
-onous	Termination for names of acids of the type $RP(OH)_2$: phosphonous acid (R = H or alkyl)
-onoyl	Termination for prefixes indicating a radical of the type $HX(O)<$ (X = P or As): phosphonoyl
-orane	Termination indicating a substituted derivative of the type XH_5: dichlorotriphenyl-phosphorane (X = P)
-oryl	Termination for prefix indicating a group of the type X(O): phosphoryl (X = P)
-ous	Termination of a name of an oxoacid of a central element in an oxidation state lower than the highest. This nomenclature is not generally recommended: phosphorous
-y	Termination for names of certain radicals containing oxygen: hydroxy, carboxy
-yl	Common termination for names of radicals: methyl, phosphanyl, uranyl
	Termination of trivial names of many oxygenated cations: uranyl
-ylene	Termination for names of bivalent radicals of the carbon family: methylene
-ylidene	Termination for name of a radical formed by the loss of two hydrogen atoms from the same atom of a chain or a ring
-ylidyne	Termination for name of radical formed by the loss of three hydrogen atoms from the same atom, usually forming a partner in a triple bond
-yne	Termination indicating the presence of a triple bond between two atoms

2. Combined affixes

-anium	Termination to names of cations derived by hydrogen cation addition to molecular hydrides with -ane-type names

TABLE IX

Table IX (*Continued*)

-ato	Termination to name of an organic or inorganic ion serving as a ligand: sulfato, acetylacetonato
-diyl	Termination indicating the loss of two hydrogen atoms from the same atom, which is then capable of forming two single bonds: phosphanediyl, HP <
-ido	Modification to name of an anion of name otherwise ending with -ide, serving as a ligand: disulfido
-inate	Termination for names of anions of certain non-carbon acids having names ending with -inic: phosphinate
-inato	Modification to the name of an -inate anion when serving as a ligand
-ito	Termination for the name of an anion deriving from an -ous acid, and serving as a ligand
-onate	Termination for names of anions deriving from an -onic acid: phosphonate
-onato	Modification to name of an -onate anion when serving as a ligand: phosphonato
-onium	Termination of name for a cation formed by addition of a hydrogen cation to a molecular hydride: phosphonium
-triyl	Termination indicating the loss of three hydrogen atoms from the same atom which is then capable of forming three single bonds: phosphanetriyl, –P <

TABLE X

Table X Representation of ligand names by abbreviations*

Abbreviation	Common name	Systematic name
Diketones		
Hacac	acetylacetone	2,4-pentanedione
Hhfa	hexafluoroacetylacetone	1,1,1,5,5,5-hexafluoro-2,4-pentanedione
Hba	benzoylacetone	1-phenyl-1,3-butanedione
Hfod	1,1,1,2,2,3,3-heptafluoro-7,7-dimethyl-4,6-octanedione	6,6,7,7,8,8,8-heptafluoro-2,2-dimethyl-3,5-octanedione
Hfta	trifluoroacetylacetone	1,1,1-trifluoro-2,4-pentanedione
Hdbm	dibenzoylmethane,	1,3-diphenyl-1,3-propanedione
Hdpm	dipivaloylmethane	2,2,6,6-tetramethyl-3,5-heptanedione
Amino alcohols		
Hea	ethanolamine	2-aminoethanol
H_3tea	triethanolamine	2,2',2''-nitrilotriethanol
H_2dea	diethanolamine	2,2'-iminodiethanol
Hydrocarbons		
cod	cyclooctadiene	1,5-cyclooctadiene
cot	cyclooctatetraene	1,3,5,7-cyclooctatetraene
Cp	cyclopentadienyl	cyclopentadienyl
Cy	cyclohexyl	cyclohexyl
Ac	acetyl	acetyl
Bu	butyl	butyl
Bzl	benzyl	benzyl
Et	ethyl	ethyl
Me	methyl	methyl
nbd	norbornadiene	bicyclo[2.2.1]hepta-2,5-diene
Ph	phenyl	phenyl
Pr	propyl	propyl
Heterocycles		
py	pyridine	pyridine
thf	tetrahydrofuran	tetrahydrofuran
Hpz	pyrazole	1*H*-pyrazole
Him	imidazole	1*H*-imidazole
terpy	2,2',2''-terpyridine	2,2':6',2''-terpyridine
picoline	α-picoline	2-methylpyridine
Hbpz$_4$	hydrogen tetra(1-pyrazolyl)borate(1−)	hydrogen tetrakis(1*H*-pyrazolato-*N*)borate(1−)
isn	isonicotinamide	4-pyridinecarboxamide
nia	nicotinamide	3-pyridinecarboxamide
pip	piperidine	piperidine

* The following practices should be followed in the use of abbreviations. It should be assumed that the reader will not be familiar with the abbreviations. Consequently, all text should explain the abbreviations it uses. The abbreviations in this table are widely used, and it is hoped that they will become standard. The commonly accepted abbreviations for organic groups (Me, methyl; Et, ethyl; Ph, phenyl, etc.) should not be used with any other meanings. The most useful abbreviations are those that readily suggest the ligand in question, either because they are obviously derived from the ligand name or because they are systematically related to structure. The sequential positions of ligand abbreviations in formulae should be in accord with Chapter I-4. Lower case letters are used for all abbreviations, except for those of certain hydrocarbon radicals. In formulae, the ligand abbreviation should be set off with parentheses, as in $[Co(en)_3]^{3+}$. Those hydrogen atoms that can be replaced by the metal atom are shown in the abbreviation by the symbol H. Thus, the molecule Hacac forms an anionic ligand that is abbreviated acac.

TABLE X

Table X (*Continued*)

lut	lutidine	2,6-dimethylpyridine
Hbim	benzimidazole	1*H*-benzimidazole

Chelating and other ligands

H_4edta	ethylenediaminetetraacetic acid	(1,2-ethanediyldinitrilo)tetraacetic acid
H_5dtpa	*N,N,N′,N″,N″*-diethylenetriaminepentaacetic acid	[[(carboxymethyl)imino]bis(1,2-ethanediyl-nitrilo)]tetraacetic acid
H_3nta	nitrilotriacetic acid	
H_4cdta	*trans*-1,2-cyclohexanediaminetetraacetic acid	*trans*-(1,2-cyclohexanediyldinitrilo)tetraacetic acid
H_2ida	iminodiacetic acid	iminodiacetic acid
dien	diethylenetriamine	*N*-(2-aminoethyl)-1,2-ethanediamine
en	ethylenediamine	1,2-ethanediamine
pn	propylenediamine	1,2-propanediamine
tmen	*N,N,N′,N′*-tetramethylethylenediamine	*N,N,N′,N′*-tetramethyl-1,2-ethanediamine
tn	trimethylenediamine	1,3-propanediamine
tren	tris(2-aminoethyl)amine	*N,N*-bis(2-aminoethyl)-1,2-ethanediamine
trien	triethylenetetramine	*N,N′*-bis(2-aminoethyl)-1,2-ethanediamine
chxn	1,2-diaminocyclohexane	1,2-cyclohexanediamine
hmta	hexamethylenetetramine	1,3,5,7-tetraazatricyclo[3.3.1.1^{3,7}]decane
Hthsc	thiosemicarbazide	hydrazinecarbothioamide
depe	1,2-bis(diethylphosphino)ethane	1,2-ethanediylbis(diethylphosphine)
diars	*o*-phenylenebis(dimethylarsine)	1,2-phenylenebis(dimethylarsine)
dppe	1,2-bis(diphenylphosphino)ethane	1,2-ethanediylbis(diphenylphosphine)
diop	2,3-*O*-isopropylidene-2,3-dihydroxy-1,4-bis(diphenylphosphino)butane	3,4-bis[(diphenylphosphinyl)methyl]-2,2-dimethyl-1,3-dioxolane
triphos		[2-[(diphenylphosphino)methyl]-2-methyl-1,3-propanediyl]bis(diphenylphosphine)
hmpa	hexamethylphosphoric triamide	hexamethylphosphoric triamide
bpy	2,2′-bipyridine	2,2′-bipyridine
H_2dmg	dimethylglyoxime	2,3-butanedione dioxime
dmso	dimethyl sulfoxide	sulfinyldimethane
phen	1,10-phenanthroline	1,10-phenanthroline
tu	thiourea	thiourea
Hbig	biguanide	imidodicarbonimidic diamide
HEt$_2$dtc	diethyldithiocarbamic acid	diethylcarbamodithioic acid
H_2mnt	maleonitriledithiol	2,3-dimercapto-2-butenedinitrile
tcne	tetracyanoethylene	ethenetetracarbonitrile
tcnq	tetracyanoquinodimethan	2,2′-(2,5-cyclohexadiene-1,4-diylidene)bis(1,3-propanedinitrile)
dabco	triethylenediamine	1,4-diazabicyclo[2.2.2]octane
2,3,2-tet	1,4,8,11-tetraazaundecane	*N,N′*-bis(2-aminoethyl)-1,3-propanediamine
3,3,3-tet	1,5,9,13-tetraazatridecane	*N,N′*-bis(3-aminopropyl)-1,3-propanediamine
ur	urea	urea
dmf	dimethylformamide	*N,N*-dimethylformamide

Schiff base

H_2salen	bis(salicylidene)ethylenediamine	2,2′-[1,2-ethanediylbis(nitrilo-methylidyne)]diphenol
H_2acacen	bis(acetylacetone)ethylenediamine	4,4′-(1,2-ethanediyldinitrilo)bis(2-pentanone)
H_2salgly	salicylideneglycine	*N*-[(2-hydroxyphenyl)methylene]glycine
H_2saltn	bis(salicylidene)-1,3-diaminopropane	2,2′-[1,3-propanediylbis(nitrilo-methylidyne)]diphenol

TABLE X

Table X (*Continued*)

H$_2$saldien	bis(salicylidene)diethylenetriamine	2,2'-[iminobis(1,2-ethanediylnitrilo-methylidyne)]diphenol
H$_2$tsalen	bis(2-mercaptobenzylidene)ethylenediamine	2,2'-[1,2-ethanediylbis(nitrilo-methylidyne)]dibenzenethiol

Macrocycles

18-crown-6	1,4,7,10,13,16-hexaoxacyclooctadecane	1,4,7,10,13,16-hexaoxacyclooctadecane
benzo-15-crown-5	2,3-benzo-1,4,7,10,13-pentaoxacyclopentadec-2-ene	2,3,5,6,8,9,11,12-octahydro-1,4,7,10,13-benzopentaoxacyclopentadecene
cryptand 222	4,7,13,16,21,24-hexaoxa-1,10-diazabicyclo[8.8.8]hexacosane	4,7,13,16,21,24-hexaoxa-1,10-diazabicyclo[8.8.8]hexacosane
cryptand 211	4,7,13,18-tetraoxa-1,10-diaza bicyclo[8.5.5]icosane	4,7,13,18-tetraoxa-1,10-diazabicyclo[8.5.5]icosane
[12]aneS$_4$	1,4,7,10-tetrathiacyclododecane	1,4,7,10-tetrathiacyclododecane
H$_2$pc	phthalocyanine	phthalocyanine
H$_2$tpp	tetraphenylporphyrin	5,10,15,20-tetraphenylporphyrin
H$_2$oep	octaethylporphyrin	2,3,7,8,12,13,17,18-octaethylporphyrin
ppIX	protoporphyrin IX	3,7,12,17-tetramethyl-8,13-divinylporphyrin-2,18-dipropanoic acid
[18]aneP$_4$O$_2$	1,10-dioxa-4,7,13,16-tetraphosphacyclo-octadecane	1,10-dioxa-4,7,13,16-tetraphosphacyclooctadecane
[14]aneN$_4$	1,4,8,11-tetraazacyclotetradecane	1,4,8,11-tetraazacyclotetradecane
[14]1,3-dieneN$_4$	1,4,8,11-tetraazacyclotetradeca-1,3-diene	1,4,8,11-tetraazacyclotetradeca-1,3-diene
Me$_4$[14]-aneN$_4$	2,3,9,10-tetramethyl-1,4,8,11-tetraazacyclo-tetradecane	2,3,9,10-tetramethyl-1,4,8,11-tetraazacyclo-tetradecane
cyclam		1,4,8,11-tetraazacyclotetradecane

Appendix

Below are listed the three most commonly published formats of the Periodic Table of the Elements: the so-called eight-column table, the eighteen-column table, and the thirty-two-column table. These three formats and other possible formats that have been proposed serve to reflect various aspects of electronic structure and chemical and physical properties, and to reflect cultural or historic development. This Commission does not wish to deprecate any specific Periodic Table format that chemists may deem necessary to interpret lucidly chemical, physical, or cultural aspects of the basic science of chemistry. However, it is true that common world-wide practice in teaching and research overwhelmingly supports the eighteen-column format.

Reference to Table A-II below emphatically illustrates the chaotic condition of Group or family labelling that existed when the Commission undertook to revise the 1970 *Nomenclature of Inorganic Chemistry*. The '1970 Nomenclature' followed an earlier precedent for the use of the letters A and B to distinguish between the left and right side families. While this labelling is in common use in large areas of the world, other equally large constituencies of the chemical community utilized the A–B labelling that distinguishes the Main Group Elements from the Transition Group Elements. In addition to these and the 1–18 column labelling used here, many other schemes have been proposed and used locally by small communities to serve their own specific needs (Note Aa).

While it is neither the intent, nor the purpose of the IUPAC Commission on the Nomenclature of Inorganic Chemistry arbitrarily to set the format of the Periodic Table to be used in all parts of the world, it is the responsibility of the Commission to offer broadly useful nomenclature proposals where direct conflicts in usage occur. After extensive discussions and many public appeals for comment, the Commission concluded that the use of the 1 to 18 numbering for the eighteen columns provides a clear and unambiguous labelling for reference within the current revision of '1970 Nomenclature'. The members of the Commission also reasoned that the 1–18 column numbering provides an alternative labelling (to the conflicting use of the A–B notation) that satisfies the need for clear and precise communication.

The Commission has decided that over whatever length of time is required by appropriate and effective process, any ultimate recommendations on the format and family labelling of the Periodic Table must be responsive to the broadest possible constituency. Each group of this larger population is invited and urged to seek and present its local consensus to the IUPAC Commission on the Nomenclature of Inorganic Chemistry, Commission II.2.

Note Aa. W. C. Fernelius and W. H. Powell, *J. Chem. Ed.*, **59**, 504 (1982); E. Fluck, *Pure Appl. Chem.*, **60**, 431 (1988); W. C. Fernelius, *J. Chem. Ed.*, **63**, 263 (1986); J. Emsley, *New Scient.*, 7 March 1985, p. 33.

Table A-I. Short-form of the Periodic Table

	I		II		III		IV		V		VI		VII		VIII			0
	A	B	A	B	A	B	A	B	A	B	A	B	A	B				
	1 H																	2 He
	3 Li		4 Be		5 B		6 C		7 N		8 O		9 F					10 Ne
	11 Na		12 Mg		13 Al		14 Si		15 P		16 S		17 Cl					18 Ar
	19 K	29 Cu	20 Ca	30 Zn	21 Sc	31 Ga	22 Ti	32 Ge	23 V	33 As	24 Cr	34 Se	25 Mn	35 Br	26 Fe	27 Co	28 Ni	36 Kr
	37 Rb	47 Ag	38 Sr	48 Cd	39 Y	49 In	40 Zr	50 Sn	41 Nb	51 Sb	42 Mo	52 Te	43 Tc	53 I	44 Ru	45 Rh	46 Pd	54 Xe
	55 Cs	79 Au	56 Ba	80 Hg	57 La*	81 Tl	72 Hf	82 Pb	73 Ta	83 Bi	74 W	84 Po	75 Re	85 At	76 Os	77 Ir	78 Pt	86 Rn
	87 Fr		88 Ra		89 Ac**													

*Including lanthanoids (57–71).
**Including actinoids (89–103).

Table A-II. Eighteen-column numbering in a conventional long-form Periodic Table

	1	2	3	4	5	6	7	8	9	10	11	12	13	14	15	16	17	18
IUPAC 1988	1	2	3	4	5	6	7	8	9	10	11	12	13	14	15	16	17	18
IUPAC 1970	IA	IIA	IIIA	IVA	VA	VIA	VIIA		VIIIA		IB	IIB	IIIB	IVB	VB	VIB	VIIB	VIIIB
Deming 1923	IA	IIA	IIIB	IVB	VB	VIB	VIIB		VIIIB		IB	IIB	IIIA	IVA	VA	VIA	VIIA	VIIIA
	1 H																	
	3 Li	4 Be											5 B	6 C	7 N	8 O	9 F	10 Ne
	11 Na	12 Mg											13 Al	14 Si	15 P	16 S	17 Cl	18 Ar
	19 K	20 Ca	21 Sc	22 Ti	23 V	24 Cr	25 Mn	26 Fe	27 Co	28 Ni	29 Cu	30 Zn	31 Ga	32 Ge	33 As	34 Se	35 Br	36 Kr
	37 Rb	38 Sr	39 Y	40 Zr	41 Nb	42 Mo	43 Tc	44 Ru	45 Rh	46 Pd	47 Ag	48 Cd	49 In	50 Sn	51 Sb	52 Te	53 I	54 Xe
	55 Cs	56 Ba	57–71 La–Lu†	72 Hf	73 Ta	74 W	75 Re	76 Os	77 Ir	78 Pt	79 Au	80 Hg	81 Tl	82 Pb	83 Bi	84 Po	85 At	86 Rn
	87 Fr	88 Ra	89–103 Ac–Lr††															
			57 †La	58 Ce	59 Pr	60 Nd	61 Pm	62 Sm	63 Eu	64 Gd	65 Tb	66 Dy	67 Ho	68 Er	69 Tm	70 Yb	71 Lu	
			89 ††Ac	90 Th	91 Pa	92 U	93 Np	94 Pu	95 Am	96 Cm	97 Bk	98 Cf	99 Es	100 Fm	101 Md	102 No	103 Lr	

Table A-III. Thirty-two-column form of the Periodic Table

1	2	3	4	5	6	7	8	9	10	11	12	13	14	15	16	17	18	IUPAC 1988
IA	IIA	IIIA	IVA	VA	VIA	VIIA	VIIIA	VIIIA		IB	IIB	IIIB	IVB	VB	VIB	VIIB	VIIIB	IUPAC 1970
IA	IIA	IIIB	IVB	VB	VIB	VIIB	VIIIB	VIIIB		IB	IIB	IIIA	IVA	VA	VIA	VIIA	VIIIA	Deming 1923
1 H																	2 He	
3 Li	4 Be											5 B	6 C	7 N	8 O	9 F	10 Ne	
11 Na	12 Mg											13 Al	14 Si	15 P	16 S	17 Cl	18 Ar	
19 K	20 Ca	21 Sc	22 Ti	23 V	24 Cr	25 Mn	26 Fe	27 Co	28 Ni	29 Cu	30 Zn	31 Ga	32 Ge	33 As	34 Se	35 Br	36 Kr	
37 Rb	38 Sr	39 Y	40 Zr	41 Nb	42 Mo	43 Tc	44 Ru	45 Rh	46 Pd	47 Ag	48 Cd	49 In	50 Sn	51 Sb	52 Te	53 I	54 Xe	
55 Cs	56 Ba	71 Lu	72 Hf	73 Ta	74 W	75 Re	76 Os	77 Ir	78 Pt	79 Au	80 Hg	81 Tl	82 Pb	83 Bi	84 Po	85 At	86 Rn	
87 Fr	88 Ra	103 Lr																

Lanthanides:

57 La	58 Ce	59 Pr	60 Nd	61 Pm	62 Sm	63 Eu	64 Gd	65 Tb	66 Dy	67 Ho	68 Er	69 Tm	70 Yb

Actinides:

89 Ac	90 Th	91 Pa	92 U	93 Np	94 Pu	95 Am	96 Cm	97 Bk	98 Cf	99 Es	100 Fm	101 Md	102 No

283

Index

INDEX

hapto symbol (η) 202–4
heteropolyacids 134–5
history of nomenclature 1–2
homopolyacids 133–4
hydrates 67
hydrides 6, 82–6
  mononuclear 83–5
  oligonuclear 85–6
  substituted derivatives 86–96
    chain compounds 88–91
    cyclic compounds 91–6
    mononuclear 87–8
  see also boron hydrides
hydrogen
  isotopes of 38
  as ligand 153
  boron hydride names giving hydrogen atom
    distribution 223–4
  oxidation number 66
  salts containing acid hydrogen 118–19
  substitution in boron hydrides 225–7
hydrogen nomenclature 125–31
hydrons 103, 104–5
hydroxides 53n, 120
  double 120–1
hyphens 17
hypho-boranes 211, 214

indexes 7–8
infinitely adaptive structures 79
infix replacement nomenclature 137–8
intercalates 79
intermetallic compounds 53
international cooperation, history of 8–9
ions 248–73
  charge on *see* charge
  polyatomic 53
  proportions in coordination compounds
    152–3
  see also anions; cations
iso- in boron hydride nomenclature 224
isopolyacids 133–4
isotopes 32, 35, 38
isotopically modified compounds 55–6
italic letters 24–5

kappa (κ) convention 98–9, 174–9
klado-boranes 211, 214
Kröger-Vink notation 72

lambda (λ) convention 84, 86
lanthanoids (lanthanides) 43
ligands 146, 153–9
  abbreviations 54–5, 158–9, 160–3, 277–9
  based on Group 15 elements 156–8
  bridging 149, 189–90, 191–2, 193–4

chalcogen-based 154–6
halogen-based 153–4
hydrogen as 153
mononuclear coordination compounds 97
number in a coordination entity 151
organic 158, 159
priorities 31–2
see also chelation
ligating atoms
  designation in polydentate ligands 173–9
  priority numbers and 168–9
locant designators 27–8, 29
  hydrides 85–6, 88
lower case letters 28

mass number 35, 37–8
metal–metal bonds 150
metallaboranes 231–2
metallacarbaboranes 231–2
metallocenes 204–6
methods of nomenclature 3–8
mineral names 70–1
minus signs 17n, 18
misfit structures 77
modulated structures 76–7
molecular formulae 45–6
monocyclic compounds 91–4
mononuclear coordination compounds 97
  with monodentate ligands 150–9
mu (μ) 50, 189–90
multiplicative prefixes 11, 27, 63–5, 243

neo- in boron hydride nomenclature 224
nido-boranes 211, 212–13, 214, 224–5
nitrosyl group 66
noble gases 43, 61n
non-stoichiometric phases 70, 76–9
nuclear reactions 38
nuclides 35
numerals, usages for 22–4
numerical prefixes 11, 27, 63–5, 243

octahedral complexes 171, 183–5, 186–8
oligonuclear coordination compounds 99–100,
  192–6
optically active compounds 49
organic ligands 158, 159
organic priority orders 31
organic radicals 66, 199–201
organoboron compounds 232
organometallic species 198–206
  metallocenes 204–6
  with unsaturated ligands 201–4
oxidation number 11, 65–6, 148–9, 151, 152–3
oxidation states 47–8, 48n
oxides 120
  double 120–1

287